# SEISMOLOGICAL RESEARCH REQUIREMENTS FOR A COMPREHENSIVE TEST-BAN MONITORING SYSTEM

Panel on Seismological Research Requirements for a Comprehensive
  Test-Ban Monitoring System
Committee on Seismology
Board on Earth Sciences and Resources
Commission on Geosciences, Environment, and Resources
National Research Council

National Academy Press
Washington, D.C. 1995

**NOTICE**: The project that is the subject of this report was approved by the Governing Board of the National Research Council, whose members are drawn from the councils of the National Academy of Sciences, the National Academy of Engineering, and the Institute of Medicine. The members of the committee responsible for the report were chosen for their special competences and with regard for appropriate balance.

This report has been reviewed by a group other than the authors according to procedures approved by a Report Review Committee consisting of members of the National Academy of Sciences, the National Academy of Engineering, and the Institute of Medicine.

Support for this study was provided by the Advanced Research Projects Agency.

**Library of Congress Catalog Card No. 95-70295**
**International Standard Book Number 0-309-05332-3**

Additional copies of this report are available from:

National Academy Press
2101 Constitution Ave., NW
Box 285
Washington, DC 20055
800-624-6242
202-334-3313 (in the Washington Metropolitan Area)

B-657

**COVER:** The cover shows a seismograph recording of a presumed French underground nuclear test in the Tuamotu Archipelago in the South Pacific Ocean. The event had a magnitude of 4.7, and it occurred at 21:30:00 GMT on September 5, 1995. The location of the event is indicated in the lower left corner. The recording is from the GERESS seismic array in Germany, noted by the point on the right of the cover. As discussed in this report, GERESS is a primary station in the GSETT-3 network that is being developed to verify a Comprehensive Test Ban Treaty. Cover figure courtesy of Charles Meade, Board on Earth Sciences and Resources, National Research Council.

Copyright ©1995 by the National Academy of Sciences.
All rights reserved.

Printed in the United States of America.

## PANEL ON SEISMOLOGICAL RESEARCH REQUIREMENTS FOR A COMPREHENSIVE TEST-BAN MONITORING SYSTEM

THORNE LAY, *Chair,* University of California, Santa Cruz
CLARENCE R. ALLEN, California Institute of Technology, Pasadena
ROBERT BLANDFORD, Air Force Technical Applications Center, Arlington, Virginia
ROBERT HAMILTON, U.S. Geological Survey, Reston, Virginia
WILLARD J. HANNON, JR., Lawrence Livermore National Laboratory, Livermore, California
THOMAS H. JORDAN, Massachusetts Institute of Technology, Cambridge
CHARLES A. LANGSTON, Pennsylvania State Univesity, University Park
PAUL G. RICHARDS, Lamont-Doherty Earth Observatory, Palisades, New York
BARBARA ROMANOWICZ, University of California, Berkeley
TERRY C. WALLACE JR., University of Arizona, Tucson

**National Research Council Staff**
WILLIAM E. BENSON, Senior Program Officer
CHARLES MEADE, Program Officer
JUDITH L. ESTEP, Administrative Assistant

## COMMITTEE ON SEISMOLOGY

THOMAS H. JORDAN, *Chair*, Massachusetts Institute of Technology, Cambridge
F. A. DAHLEN, Princeton University, New Jersey
STEVEN M. DAY, San Diego State University, California
THOMAS C. HANKS, U.S. Geological Survey, Menlo Park, California
CHARLES A. LANGSTON, Pennsylvania State University, University Park
THORNE LAY, University of California, Santa Cruz
STEWART A. LEVIN, Mobil Exploration & Production Technical Center, Dallas, Texas
STEPHEN D. MALONE, University of Washington, Seattle
JAMES R. RICE, Harvard University, Cambridge, Massachusetts
PAUL G. SOMERVILLE, Woodward-Clyde Consultants, Pasadena, California
ANNE M. TREHU, Oregon State University, Corvallis

**National Research Council Staff**
WILLIAM E. BENSON, Senior Program Officer
CHARLES MEADE, Program Officer
JUDITH L. ESTEP, Administrative Assistant

# BOARD ON EARTH SCIENCES AND RESOURCES

J. FREEMAN GILBERT, *Chair,* University of California, San Diego
GAIL M. ASHLEY, Rutgers University, Piscataway, New Jersey
THURE CERLING, University of Utah, Salt Lake City
MARK P. CLOOS, University of Texas at Austin
WILLIAM R. DICKINSON, University of Arizona, Tucson
JOEL DARMSTADTER, Resources for the Future, Washington, D.C.
MARCO T. EINAUDI, Stanford University, California
NORMAN H. FOSTER, Independent Petroleum Geologist, Denver, Colorado
CHARLES G. GROAT, Louisiana State University, Baton Rouge
DONALD C. HANEY, University of Kentucky, Lexington
SUSAN M. KIDWELL, University of Chicago, Illinois
PHILIP E. LaMOREAUX, P.E. LaMoreaux and Associates, Inc., Tuscaloosa, Alabama
SUSAN M. LANDON, Thomasson Partner Associates, Denver, Colorado
J. BERNARD MINSTER, University of California, San Diego
ALEXANDRA NAVROTSKY, Princeton University, New Jersey
JILL D. PASTERIS, Washington University, St. Louis, Missouri
EDWARD C. ROY JR., Trinity University, San Antonio, Texas

**National Research Council Staff**
CRAIG M. SCHIFFRIES, Staff Director
THOMAS M. USSELMAN, Associate Staff Director
INA B. ALTERMAN, Senior Program Officer
WILLIAM E. BENSON, Senior Program Officer
KEVIN D. CROWLEY, Senior Program Officer
ANNE M. LINN, Senior Program Officer
CHARLES MEADE, Program Officer
LALLY A. ANDERSON, Staff Associate
VERNA J. BOWEN, Administrative Assistant
JENNIFER T. ESTEP, Administrative Assistant
JUDITH L. ESTEP, Administrative Assistant

# COMMISSION ON GEOSCIENCES, ENVIRONMENT, AND RESOURCES

M. GORDON WOLMAN, *Chair*, The Johns Hopkins University, Baltimore, Maryland
PATRICK R. ATKINS, Aluminum Company of America, Pittsburgh, Pennsylvania
JAMES P. BRUCE, Canadian Climate Program Board, Ottawa, Ontario, Canada
WILLIAM L. FISHER, The University of Texas, Austin
GEORGE M. HORNBERGER, University of Virginia, Charlottesville
DEBRA KNOPMAN, Progressive Policy Institute, Washington, D.C.
PERRY L. McCARTY, Stanford University, California
JUDY McDOWELL, Woods Hole Oceanographic Institution, Massachusetts
S. GEORGE PHILANDER, Princeton University, New Jersey
RAYMOND A. PRICE, Queen's University at Kingston, Ontario, Canada
THOMAS A. SCHELLING, University of Maryland, College Park
ELLEN K. SILBERGELD, University of Maryland Medical School, Baltimore, Maryland
STEVEN M. STANLEY, The Johns Hopkins University, Baltimore, Maryland
VICTORIA J. TSCHINKEL, Landers and Parsons, Tallahassee, Florida

**National Research Council Staff**
STEPHEN RATTIEN, Executive Director
STEPHEN D. PARKER, Associate Executive Director
MORGAN GOPNIK, Assistant Executive Director
GREGORY SYMMES, Reports Officer
JIM MALLORY, Administrative Officer
SANDI FITZPATRICK, Administrative Associate
SUSAN SHERWIN, Project Assistant

# PREFACE

At the request of the Advanced Research Projects Agency (ARPA), the Committee on Seismology of the National Research Council established the Panel on Seismological Research Requirements for a Comprehensive Test-Ban Monitoring System to address issues associated with establishing an International Seismic Monitoring System (ISMS) for verifying a Comprehensive Test-Ban Treaty (CTBT). Major decisions are now being made at an international level that will affect seismological monitoring and research efforts for the next few decades. A global network of high-quality seismic arrays and broadband stations will provide data to the ISMS, with participating states having access to the data for national treaty verification functions. The ISMS data can be used to augment both earthquake monitoring and basic earth science research capabilities in the United States, as long as the data characteristics are adequate and the data are readily available to the broad seismological community in a timely manner. Issues considered in this report include specifications of ISMS instrumentation, mechanisms that must be established to provide general access to ISMS data, and the U.S. research infrastructure needed to support the ISMS and national verification functions. This report provides recommendations on both specific technical issues and broader policy issues related to U.S. participation in the new monitoring system. The recommendations are organized under three specific charges to the panel, which are presented in full in Appendix A.

The National Academy of Sciences is a private, nonprofit, self-perpetuating society of distinguished scholars engaged in scientific and engineering research, dedicated to the furtherance of science and technology and to their use for the general welfare. Upon the authority of the charter granted to it by the Congress in 1863, the Academy has a mandate that requires it to advise the federal government on scientific and technical matters. Dr. Bruce Alberts is president of the National Academy of Sciences.

The National Academy of Engineering was established in 1964, under the charter of the National Academy of Sciences, as a parallel organization of outstanding engineers. It is autonomous in its administration and in the selection of its members, sharing with the National Academy of Sciences the responsibility for advising the federal government. The National Academy of Engineering also sponsors engineering programs aimed at meeting national needs, encourages education and research, and recognizes the superior achievements of engineers. Dr. Harold Liebowitz is president of the National Academy of Engineering.

The Institute of Medicine was established in 1970 by the National Academy of Sciences to secure the services of eminent members of appropriate professions in the examination of policy matters pertaining to the health of the public. The Institute acts under the responsibility given to the National Academy of Sciences by its congressional charter to be an adviser to the federal government and, upon its own initiative, to identify issues of medical care, research, and education. Dr. Kenneth I. Shine is president of the Institute of Medicine.

The National Research Council was organized by the National Academy of Sciences in 1916 to associate the broad community of science and technology with the Academy's purposes of furthering knowledge and advising the federal government. Functioning in accordance with general policies determined by the Academy, the Council has become the principal operating agency of both the National Academy of Sciences and the National Academy of Engineering in providing services to the government, the public, and the scientific and engineering communities. The Council is administered jointly by both Academies and the Institute of Medicine. Dr. Bruce Alberts and Dr. Harold Liebowitz are chairman and vice-chairman, respectively, of the National Research Council.

# TABLE OF CONTENTS

**1 EXECUTIVE SUMMARY**   1
    Recommendations, 3
        Data Characteristics, 3
        Data Access Within the United States, 4
        Research Feedback, 6

**2 INTRODUCTION**   9
    Planned International Seismic Monitoring System, 10
    Existing Seismological Systems, 13

**3 ISMS DATA CHARACTERISTICS**   23
    Introduction and Background, 23
    Discussion of ISMS Station Technical Requirements, 25
    Desired Raw and Processed ISMS Data Streams, 31

**4 DISTRIBUTION OF ISMS DATA WITHIN THE UNITED STATES**   33
    Introduction and Background, 34
        Agencies with an Interest in Seismic Data, 34
        Seismic Waveform Data, 35
        Seismological Event Bulletins, 38
    Fundamental Guidelines for Data Access Issues, 42
    Agency-Specific Recommendations Concerning Data Access, 46

**5 ISMS AND U.S. NATIONAL VERIFICATION RESEARCH AND DEVELOPMENT INFRASTRUCTURE**   53
    Introduction and Background, 54
    U.S. Research and Development Infrastructure, 59

**REFERENCES**   69

**APPENDIX A: CHARGE TO THE PANEL**   71

**APPENDIX B: RESEARCH TOPICS FOR CTBT SEISMIC MONITORING**   73

**APPENDIX C: ACRONYM LIST**   79

# LIST OF FIGURES

| | | |
|---|---|---|
| 2.1 | Conceptual model for the International Seismic Monitoring System | 12 |
| 2.2 | Schematic of the current structure of the U.S. treaty verification, earthquake monitoring, and basic research seismological efforts. | 16 |
| 2.3 | Distribution of GSETT-3 primary stations | 18 |
| 2.4 | Distribution of stations of U.S. Geological Survey regional networks | 19 |
| 2.5 | Distribution of global seismic stations contributing to ISC | 20 |
| 2.6 | Distribution of stations of the Federation of Digital Seismographic Networks | 22 |

# SEISMOLOGICAL RESEARCH REQUIREMENTS FOR A COMPREHENSIVE TEST-BAN MONITORING SYSTEM

# 1

# EXECUTIVE SUMMARY

Negotiations of a Comprehensive Test-Ban Treaty (CTBT) are now underway, and the Non-Proliferation Treaty was extended indefinitely in May 1995. Both of these are important steps in the reduction of the worldwide threat of nuclear weapons. These treaties create a need to monitor for nuclear explosions in the context of national and international efforts in nuclear arms control. Seismology, a discipline that provides the principal technology for detecting, locating, and identifying underground nuclear explosions on a global basis, is confronted with the massive new challenge of monitoring a global ban on all nuclear testing. With seismology playing a prominent role in U.S. and international treaty monitoring procedures, it is essential to plan carefully the seismological monitoring system at all levels, from the basic research programs that support the monitoring effort, to the instrumentation, to the use of the results in the national verification system. This report will address many of the key issues associated with implementing the seismological monitoring system.

The United States is now in a time of pivotal decision-making, with major issues being decided that will affect the field of seismology for the next few decades. Major expenditures by the United States and other nations are now being made to provide the seismic recording and analysis capabilities essential for a cooperative international monitoring effort. In the rapidly evolving political landscape surrounding nuclear test-ban and nonproliferation treaties, there is a window of opportunity to ensure that the international seismic system will contribute broadly to multiple issues of national concern, including earthquake monitoring and basic research on earth structure and processes, as well as treaty verification functions. Small nuclear tests, such as might be part of a clandestine weapons program, produce ground vibration levels equivalent to those of thousands of natural seismic events that occur each year. And improved seismological methods will be needed to assess the nature of these sources. The vast majority of recorded events will be natural earthquakes, and the seismic recordings made for monitoring purposes will be useful for further scientific analysis and hazard assessment.

Both broadband and short-period array data will be collected by the international treaty monitoring system, and all these data have multiple potential applications. The large quantity of both types of data offers a significant increase in the number of timely signals that can be accessed from stations around the world for earthquake monitoring and basic research applications as well as for basic monitoring applications. Rapid, widespread access to the treaty monitoring data will provide improved determination of earthquake fault mechanisms and more reliable rapid earthquake assessment and tsunami warning capabilities.

This report describes ways of ensuring the multiple use of the seismic data collected by the new treaty monitoring system, along with measures needed to sustain the treaty monitoring capabilities of the United States into the future. The recommendations address issues associated with the characteristics of the instrumentation of the international seismic monitoring system (ISMS), the critical importance of open access to the data collected by the system, and the U.S. infrastructure needed to sustain the long-term monitoring of nuclear testing treaties.

The treaty monitoring data will be of very high quality but will constitute only a fraction of the total seismic data required for earthquake monitoring and basic research. The new international seismological system that is being developed presents an opportunity to break down past barriers to broad usage of data collected by treaty monitoring activities, to the benefit of all applications. The key to achieving this goal lies in the definition of the functions of the U.S. National Data Center, which will support both the international monitoring program and the national verification function. If the monitoring capability is to be maximized and other nationally important applications are to benefit, the U.S. National Data Center mission statement must include a data access obligation and appropriate funding must be identified to support this activity. Both the archiving and the distribution of the ISMS data have cost implications for the U.S. NDC. Because no specific plan has yet been put forth, the panel did not attempt any detailed cost analyses. We have suggested what appears to be the most economical approach.

History has repeatedly demonstrated that basic seismological research efforts are an essential part of the national strategy for long-term treaty verification. These are required both to enhance treaty monitoring capabilities and to ensure a pool of seismological expertise for future monitoring efforts.

The research community can also play a part in the confidence-building process that is an essential element in the justification of the ISMS. These researchers will be advisors to their governments and will provide important independent checks and balances on the operations of the monitoring system, and sources of insight into the geophysical properties of regions of Earth, the nature of specific events of interest, and monitoring methods in general. In addition, the broader the user community is, the better the feedback about quality control issues and instrumentation problems. Such

# EXECUTIVE SUMMARY

problems are often revealed in the course of analysis of recordings for large earthquakes, which may be ignored in the national verification effort.

The specific recommendations are listed in the next section. Those concerned with Data Characteristics and Data Access have been issued, essentially in their present form in preliminary reports designed to provide timely information and assistance to the U.S. negotiating team in Geneva.

## Recommendations

Continuous recordings from the high quality, globally distributed seismometers of the ISMS can be used beneficially for numerous purposes, if the seismological system has certain attributes. These include the recording system characteristics, as discussed in Chapter 3; the availability of the data to diverse seismological communities, as discussed in Chapter 4; and a strong seismic research and development program, as discussed in Chapter 5. The large international investments in the new ISMS must not be underutilized by the United States, as has often been the case with data collected for nuclear test monitoring in the past. Relatively low-cost efforts can ensure maximum utilization of the data for a variety of activities in the national interest, as well as augment the research and development efforts that support U.S. treaty verification capabilities. The panel has addressed both specific technical issues and larger-scale infrastructure questions in pursuit of optimization of use of the ISMS data. The recommendations in this report have been framed to enhance U.S. activities in both nuclear test-ban monitoring and earthquake monitoring. Failure to follow through on the recommendations, especially those concerned with data access, will lead to duplication of effort in the seismological system and underutilization of seismic data acquired at substantial cost.

The primary recommendations of this report are summarized below:

### Data Characteristics

It is important that the data characteristics of the new ISMS stations be compatible with the broad needs of seismology in general as well as fulfilling treaty monitoring requirements. The panel's main recommendations for data characteristics involve bandwidth and recording-system specifications. The interest in high-frequency signals from small events for CTBT monitoring has led to an emphasis on that part of the seismic spectrum in the ISMS station design, but it is technologically straightforward to simultaneously record lower- frequency signals that are of primary value for earthquake monitoring and basic research on earthquake processes and earth structure. The extended bandwidth also has important potential applications in discriminating explosion and earthquake signals. Care must be taken to ensure that lower-frequency signals are not clipped when the high-frequency signals are emphasized. This involves modest enhancement of ISMS station designs, with no reduction in high-frequency capabilities. The primary recommendations from Chapter 3 are technical in nature and are given below:

- Wherever possible, without degrading the ISMS's monitoring performance, extend the bandpass of the ISMS broadband three-component elements to as low as 0.003 Hz.
- Relax the low noise requirement to the 10–20 Hz range.
- Re-evaluate the sample rate requirements.
- Relax the resolution requirements for broadband three-component elements and base the noise floor on local conditions.
- Provide better specification of the sensitivity goals, emphasizing performance at higher frequencies.
- Specify the frequency band of the system noise requirement.
- Develop a mechanism to provide data in SEED (Standard for Exchange of Earthquake Data) format in addition to other formats that might be used.
- Reconsider the data frame length requirement.
- Establish separate data availability requirements for primary and auxiliary stations.
- Relax the orientation tolerance for primary station instrumentation.

**Data Access Within the United States**

Given suitable data characteristics, the ISMS data set can contribute to diverse efforts that address earthquake monitoring and basic research on earthquakes and earth structure, as well as the nuclear test-ban monitoring effort. To enable these multiple uses of the seismic data, it is important to establish convenient pathways for data access in the United States that do not interfere with the nation's primary operations of the nuclear test-ban monitoring effort. This report proposes cost-effective strategies that will provide these pathways. The key element is to ensure that the U.S. nuclear monitoring effort and the existing data archival and distribution capabilities are integrated for the mutual benefit of all seismological applications serving the nation. The primary recommendations from Chapter 4 concern policy on data access and are given below:

- At a minimum, the development of the ISMS should augment, not reduce, the capabilities of the U.S. scientific community. Therefore, it should not restrict current paths of access to existing stations nor limit access to new unclassified stations. Implementing this guideline will require attention to preexisting international relationships, treaty language, and agreements regarding seismic data exchange.
- The U.S. position should be that the entire ISMS seismic data set should be available in a timely manner and that these data should be unclassified. Distribution within any country would of course be the responsibility of that country's National Data Center. Therefore the U.S. government should ensure that these data are readily accessible in the United States.
- The U.S. ISMS National Data Center (ISMS-NDC) is expected to receive all of the ISMS primary-network data for U.S. treaty monitoring use. The panel recommends that the U.S. ISMS-NDC should be operated under a policy that requires it to provide the U.S. scientific, disaster prevention, and earthquake monitoring communities with stable, timely access to all signals and seismic event data that it receives from the ISMS. Costs of operating the ISMS-NDC should be provided by the nuclear monitoring community; incremental system costs for external data transmission should be provided by the earthquake monitoring agencies and by agencies supporting research on nuclear explosion and earthquake monitoring. To facilitate interagency data transmission and to deal with cost issues, the ISMS-NDC should establish a multiagency advisory committee, with representation from the nuclear monitoring, earthquake monitoring, and basic research communities, to address data distribution issues.
- All broadband data from primary and auxiliary stations received by the ISMS-NDC should be made available to the earthquake monitoring agencies in the United States in near real time (possibly by direct rebroadcast from the ISMS-NDC or by satellite downlink). These data should be archived in and made accessible on various media through the Incorporated Research Institutions for Seismology's Data Management System (IRIS-DMS). This system has extensive capabilities for servicing diverse data requests and a willingness to distribute ISMS broadband data along with other global broadband seismic data. This approach provides a permanent on-line archive of the broadband ISMS data set, facilitates user access to the data, and greatly reduces the data-distribution load on the ISMS-NDC. Assuming the data are accompanied by quality-control information, the incremental costs involved should be borne by the earthquake monitoring agencies and by agencies supporting research on nuclear explosion and earthquake monitoring.
- The continuous data from auxiliary stations (most of which will not be accessed routinely by the ISMS) should continue to be archived and distributed through existing procedures of the Federation of Digital Seismographic Networks (FDSN).

Operational support for U.S. auxiliary stations should be shared by the nuclear monitoring, earthquake monitoring, and basic research agencies.

- Continuous data from short-period arrays will comprise most of the ISMS data. These data will be important for nuclear monitoring operations. Currently, the earthquake monitoring and basic research programs have limited demand for array data, but this will almost certainly grow with time. The research that supports nuclear monitoring will require access to these data. The ISMS-NDC will archive the array data, and it is certainly not cost effective to duplicate this archive. Therefore, a user-friendly interface should be established to provide access to the entire data set. We propose that the U.S. Geological Survey and/or IRIS are logical entities to coordinate with the ISMS-NDC to develop a user-friendly pathway to all of the array data. The incremental costs involved in establishing and maintaining this pathway should be borne by agencies supporting research on nuclear explosion and earthquake monitoring.

- Seismic event data (arrival times, amplitudes, ray parameters, final event bulletins) generated by the ISMS should be made available through appropriate National Data Centers to the U.S. earthquake monitoring agencies as well as to the International Seismological Centre (ISC) to enable improvements in the seismicity bulletins produced by those agencies. Electronic transmission should minimize the costs.

- The Group of Scientific Experts Technical Test #3 (GSETT-3) experiment can be used to develop and test the data distribution pathways recommended above. The data from GSETT-3 currently being collected by the ISMS-IDC can be sent directly to the USGS from the ISMS-IDC until such times as it is possible to transmit continuous data from the ISMS-IDC.

**Research Feedback**

Monitoring compliance with a CTBT poses many unprecedented technical and scientific challenges, and there will be a continuing need for basic and applied research, as well as advanced technology and automated systems development, in all of the disciplines that contribute to the monitoring system (OTA, 1988). It is especially important that the use of comparatively new technologies such as Synthetic Aperture Radar (SAR) and the Global Positioning System (GPS) be considered for integration into the base data that will continue to come from continuous seismic recording. It is essential to sustain basic research activities that will train the next generation of seismological experts vital to long-term treaty monitoring. Furthermore, it is critical to have effective means by which basic research developments are carried out, the results are tested in operational settings, and useful, cost-effective advances are implemented in the operational system. This holds for both the ISMS and the U.S. monitoring systems. Chapter 5 considers this topic in detail. It is assumed that the mission for support of monitoring research will continue to reside within the Department of Defense (DOD) and the Department of Energy (DOE), with supporting activities by the USGS and seismological research community. If these agency roles change, the basic seismological research effort must be maintained by those responsible for the functions of monitoring, verification, and hazard reporting. The primary recommendations from Chapter 5 concern management issues and are given below:

- The DOD and DOE both have valuable assets and experience that can contribute to the seismic research and development program supporting CTBT monitoring. Continuation of the current coordinated research effort is in the best interest of the United States. The overall research effort of the DOD and DOE programs should be overseen by an advisory group that addresses both research coordination and relevance. This advisory group should have access to policy-level management.
- The DOD research and development effort in support of monitoring a CTBT should have a balanced program involving basic research, exploratory development, and advanced development efforts (the standard 6.1, 6.2, 6.3 categories of DOD research efforts), and an innovative technologies effort (traditionally the role of the Advanced Research Projects Agency) servicing the end-user, which is currently the Air Force Technical Applications Center (AFTAC).
- The Air Force basic research (6.1) program in seismology, currently administered by the Air Force Office of Scientific Research (AFOSR), should be sustained, possibly with some short-term expansion, to maintain an influx of researchers and fundamental research on long-term problems associated with seismological monitoring of a CTBT.
- The Air Force exploratory development (6.2) program in seismology, currently administered by the Air Force Phillips Laboratory, should be provided with a stable base for external funding to enable effective development and transfer of promising research and technologies from the AFOSR basic research program to the Air Force operational environment.
- The Air Force advanced development research (6.3) program in seismology currently administered by AFTAC should be sustained.
- The development of the prototype ISMS International Data Center and other advanced computer technology capabilities and high risk/high return research topics currently sponsored by ARPA should be sustained.
- The DOE research and development effort in support of seismic monitoring of a CTBT should sustain its directed research program, involving national laboratory and externally funded seismic research of direct relevance to the end-user, which is currently AFTAC.

- A knowledgeable, responsible advisory mechanism should oversee the combined DOD/DOE research effort to ensure relevance and continued coordination of the programs.
- Improved communication between and among the DOD operational units and researchers in the basic and exploratory development programs is essential. Release of information about operational methodologies and procedures, lists of problem events, and comparisons of seismic bulletins from different communities are among the activities that could enhance responsiveness of the research community to the operational requirements. Communication across the various elements of the monitoring and research communities should be fostered by symposia, workshops, site visits, and advisory panels. Focused experiments, involving broad communities, should be conducted to concentrate effort on important issues.
- To the extent possible and consistent with national security considerations, an unclassified experimental test bed facility that replicates the basic U.S. and ISMS analysis procedures should be established and made broadly available to enable new developments to be tested in a realistic environment, enhancing transfer of applied research results into the operational systems.
- A research data base of important seismic recordings should be assembled and maintained. Ground truth data bases should be provided to the test bed to assess performance of new methods. A results data base and literature guide should also be established.
- Major research efforts that have potential benefits for both nuclear test and earthquake monitoring, such as enhanced association algorithms, new regional event location procedures, and event location procedures in three-dimensional models should be coordinated through interagency working groups (for example, bridging between AFTAC and the USGS, which conducts earthquake monitoring).
- A program in which postdoctoral fellows and visiting researchers are able to work at the International Data Center, as well as the U.S. National Data Center, would provide effective communication between the operational and research environments.

Implementing the recommendations of this report regarding data characteristics, data distribution, and research infrastructure will ensure that the United States derives maximum benefit from its participation in the ISMS. Optimal multiple use of the seismic data streams for nuclear test treaty monitoring, earthquake monitoring, and basic earth science research will be enabled. In addition, U.S. treaty monitoring efforts will continue to have the critical influx of research innovations, technical developments, and personnel vital to an effective monitoring operation.

# 2

# INTRODUCTION

In 1994, the National Research Council convened the Panel on Seismological Research Requirements for a Comprehensive Test-Ban Monitoring System (hereinafter, the panel) to examine issues associated with establishing an International Seismic Monitoring System (ISMS) for verifying a Comprehensive Test-Ban Treaty (CTBT). Negotiation of such a treaty is currently underway within the Conference on Disarmament (CD), with prototype versions of the ISMS being explored in a series of technical tests organized by the Group of Scientific Experts (GSE). The latest technical test, GSETT-3, commenced January 1, 1995, and may phase into the long-term operational effort of the ISMS.

While various technologies, including seismology, are essential for monitoring atmospheric and underwater explosions, seismology provides the primary means for monitoring underground nuclear explosions. In many cases, seismic waves from buried explosions can be recorded by global networks of seismometers, and the signals used to detect, locate, and identify the source of the disturbance (allowing nuclear explosions to be distinguished from conventional chemical explosions or natural earthquakes).

The seismological component of the CTBT monitoring system being considered within the CD includes the acquisition and processing of seismic data from high-quality stations and provision of the data to participating states to assist them in their national verification functions. The Advanced Research Projects Agency (ARPA) has requested advice from the National Research Council (NRC) on how the data from the CTBT monitoring system might best benefit the broader seismological community.

The NRC panel has been charged with considering the specific data characteristics desired by the broad seismological community, the procedures for providing general access to the ISMS data, and the nature of the research infrastructure that could best support the United States' ability to perform CTBT[1] monitoring. It should be noted that the topics encompassed by this charge differ in nature. (1) The recommendations regarding instrumentation characteristics are intended for technical specialists. (2) The recommendations regarding data access involve policy issues for the U.S. National Data

---

[1] The specific charges to the panel are given in full in Appendix A.

Center, U.S. government agencies involved in CTBT, and the treaty verification community in general. (3) The research infrastructure recommendations involve the federal agencies that support treaty monitoring research.

To address this broad range of issues, the panel was constituted with expertise from the nuclear monitoring, earthquake monitoring, and basic seismological research arenas. The panel obtained extensive technical advice from its affiliated members for each of the tasks, along with soliciting additional input from many seismological experts for each of the different topics. Two preliminary reports, addressing the first two charges, were produced and distributed in response to deadlines for the GSE activities related to GSETT-3 and associated planning for the final ISMS. This report provides the panel's full response to all three tasks. Detailed discussion of each task is presented in chapters 3, 4, and 5.

## Planned International Seismic Monitoring System

To provide a context for considering the three charges before the panel, this chapter outlines the current plans for the ISMS. (A prototype ISMS began operation during GSETT-3, which commenced January 1, 1995.) This chapter also presents an overview of the existing U.S. operational capabilities associated with nuclear monitoring, earthquake monitoring, and basic research activities.

The CTBT negotiations are in progress, and the ISMS model will evolve. A recent concept for the ISMS is illustrated in Figure 2.1 (from Arms Control and Nonproliferation Technologies, Second Quarter, 1994, p. 11). This system is focused on nuclear monitoring and is neither designed nor intended to replace any existing international efforts for earthquake monitoring or data acquisition for basic science applications.

The current scenario for the ISMS envisions that two main categories of seismic waveform data will flow into the system. The first comprises continuously telemetered data from primary stations, many of which will be short-period arrays and all of which will have at least one broadband three-component seismometer. The second category of data will involve auxiliary stations, all equipped with a broadband three-component sensor with on-demand, dial-up access. Only segmented time windows are expected to be retrieved from auxiliary stations by the ISMS. Many, if not all, of the auxiliary stations will be drawn from existing global seismographic networks, which currently have procedures for accessing and archiving their continuous data. All ISMS stations will have very high quality-control and maintenance requirements. A final category of supplemental data that may be provided to the ISMS involves regional bulletins,

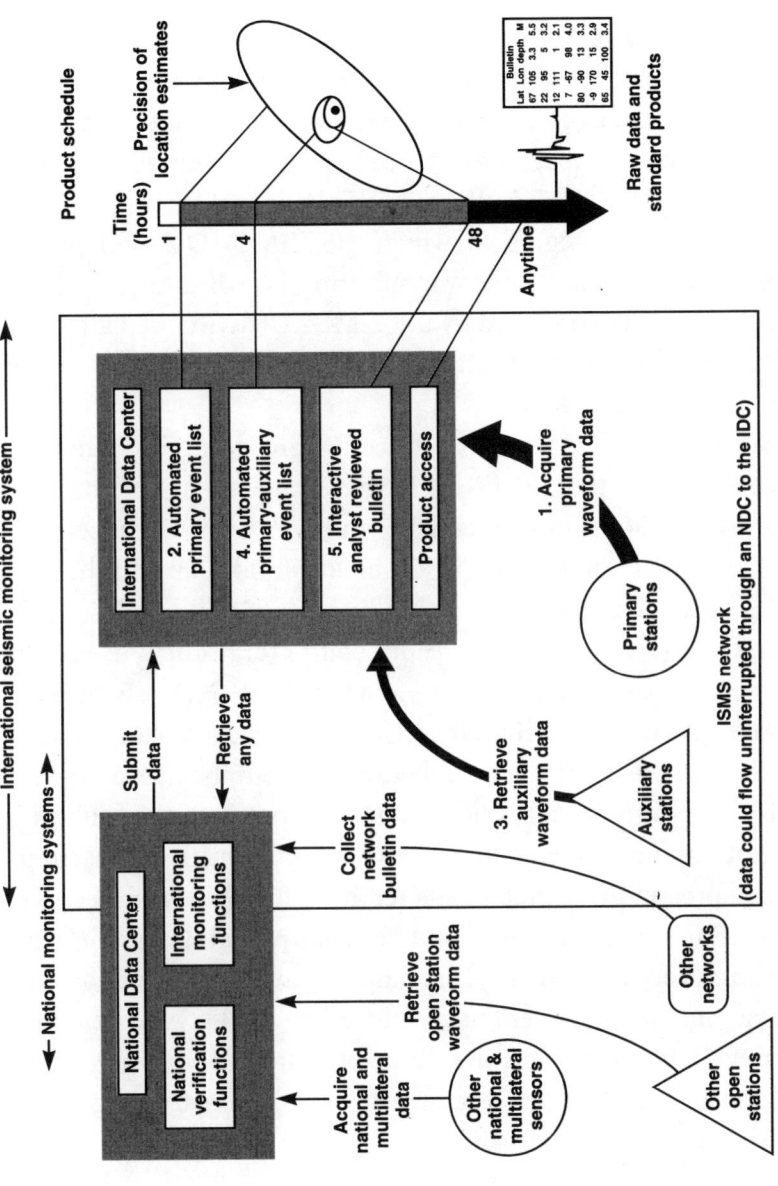

Figure 2.1. A model of the general flow of data and information between the International Seismic Monitoring System (ISMS) network stations, ISMS-IDC, and ISMS-NDCs within the International Seismic Monitoring System and within national components (left) that are external to the international system. The information on the right shows the relative schedule of products from the IDC and the improvement of the product quality, represented by the decreasing area of location confidence ellipses, as processing proceeds. Automated products from the IDC are available relatively rapidly, with human-reviewed products, in which more confidence can be placed, available at a later time. Note that no outflow of data to the research community is shown in this model (from Arms Control and Nonproliferation Technologies, Second Quarter, 1994, p. 11).

parameter data such as arrival times and amplitudes of various seismic waves, and possibly even waveform data from regional seismographic networks and other sources.

An ISMS International Data Center (ISMS-IDC) will receive data from the network of primary stations and use them to produce an automated event list within about one hour of the event. Based on this list, additional data from auxiliary stations will be accessed as needed to refine the event list within 4 hours. Analysts will review the upgraded primary-auxiliary event list and produce a final ISMS bulletin within 2 days of the end of the day of the event. The rapid preparation of this bulletin precludes incorporation of many seismic observations acquired by international earthquake monitoring efforts, so the ISMS-IDC bulletin will not be definitive with respect to global earthquake activity. The degree to which the ISMS-IDC will pursue event identification efforts related to nuclear event monitoring is still unresolved. All of the seismic data and event parameters obtained by the ISMS-IDC will be available to ISMS National Data Centers (ISMS-NDCs), which can utilize this information in independent national verification functions.

Each ISMS-NDC may have responsibilities for providing its nation's primary and auxiliary station data to the ISMS-IDC, retrieving seismic data and event parameters from the ISMS-IDC, and servicing internal verification functions. For the United States, it is likely that some of the additional national and multilateral data, combined with ISMS data in national verification functions, will be classified, as will the final nuclear monitoring event list. As a result, computer security issues will exist at the interface between the classified operations and the ISMS-NDC. In addition, there will be a need to ensure data validity within the ISMS-IDC-NDC system. The event list produced by the U.S. national verification function will emphasize identification of possible nuclear explosion signals and is not intended to produce the highest possible quality event list of earthquakes. Indeed, for events readily identified as earthquakes on the basis of location, depth, and/or signal character, no effort will be made to optimize the event parameters. For small, shallow events in continental areas, the verification event list is likely to be of very high quality, presumably superior to the event list of the ISMS-IDC. In the past, the national event list produced by the U.S. nuclear monitoring system has not been available to the unclassified community. It appears unlikely that this will change, as long as classified data streams are used in constructing the event list, even if unclassified data play the major role.

The organizational structure of the U.S. ISMS-NDC and oversight responsibilities are still unresolved, as is the issue of whether the classified national verification function will be physically separate or collocated with the ISMS-NDC. (Although no decision has been made, it seems probable that AFTAC will continue its role and be the operator of the U.S. ISMS-NDC.) The model shown in Figure 2.1 places the national verification function under the ISMS-NDC, but this is not a required structure. This

# INTRODUCTION

report will address some of the functionalities of the ISMS-NDC with respect to data archival and distribution.

This ISMS concept is being tested under the ongoing GSETT-3 experiment. The prototype ISMS-IDC is located at the Center for Monitoring Research in Arlington, Virginia, and is operated by ARPA. The prototype U.S. ISMS-NDC is operated by the Air Force Technical Applications Center (AFTAC) at Patrick Air Force Base in Florida. AFTAC will combine ISMS data with data from additional National Technical Means (NTM) in the construction of its classified event list. The USGS has a functional role in the data flow to the ISMS-NDC for GSETT-3, contributing seismic data streams that comprise much of the U.S. component of GSETT-3.

Note that there is no specific pathway for data distribution from the ISMS-NDC model in Figure 2.1. However, the GSE is presently considering its policies with respect to external data distribution in the GSETT-3. It is broadly recognized that providing access to the data is highly desirable. The panel views data distribution as an essential function to include in GSETT-3, in order to evaluate data distribution mechanisms for the future ISMS. Therefore, Chapter 4 of this report identifies possible pathways by which the unclassified seismic data and event parameters from the U.S. ISMS-NDC can be made available for other efforts related to nuclear test monitoring, earthquake studies, and emergency response. We now review existing operational capabilities and functions of different elements in the seismological systems supporting nuclear test and earthquake monitoring.

## Existing Seismological Systems

The current nuclear monitoring seismic system in the United States (Figure 2.2) is largely a classified operation, with seismic arrays in the U.S. Atomic Energy Detection Systems (USAEDS) providing data in real-time to AFTAC. The entire system involves data acquisition, data archival, and data processing, but no data distribution. A classified event bulletin with source-type discrimination and yield estimation for suspected nuclear tests has been the primary product of this nuclear monitoring system. This has been an almost entirely closed system, with limited external access to the data used in nuclear monitoring operations, even when USAEDS data have been declassified. In part, this is in compliance with bilateral agreements with the host countries for USAEDS facilities, but even some unclassified data with no such restrictions have not been available. This restricted access has precluded incorporation of the high-quality seismological data from the nuclear monitoring arena into other national efforts involving earthquake monitoring, research on earthquakes and earth structure, and even research on nuclear monitoring. The Air Force does not have any responsibility to

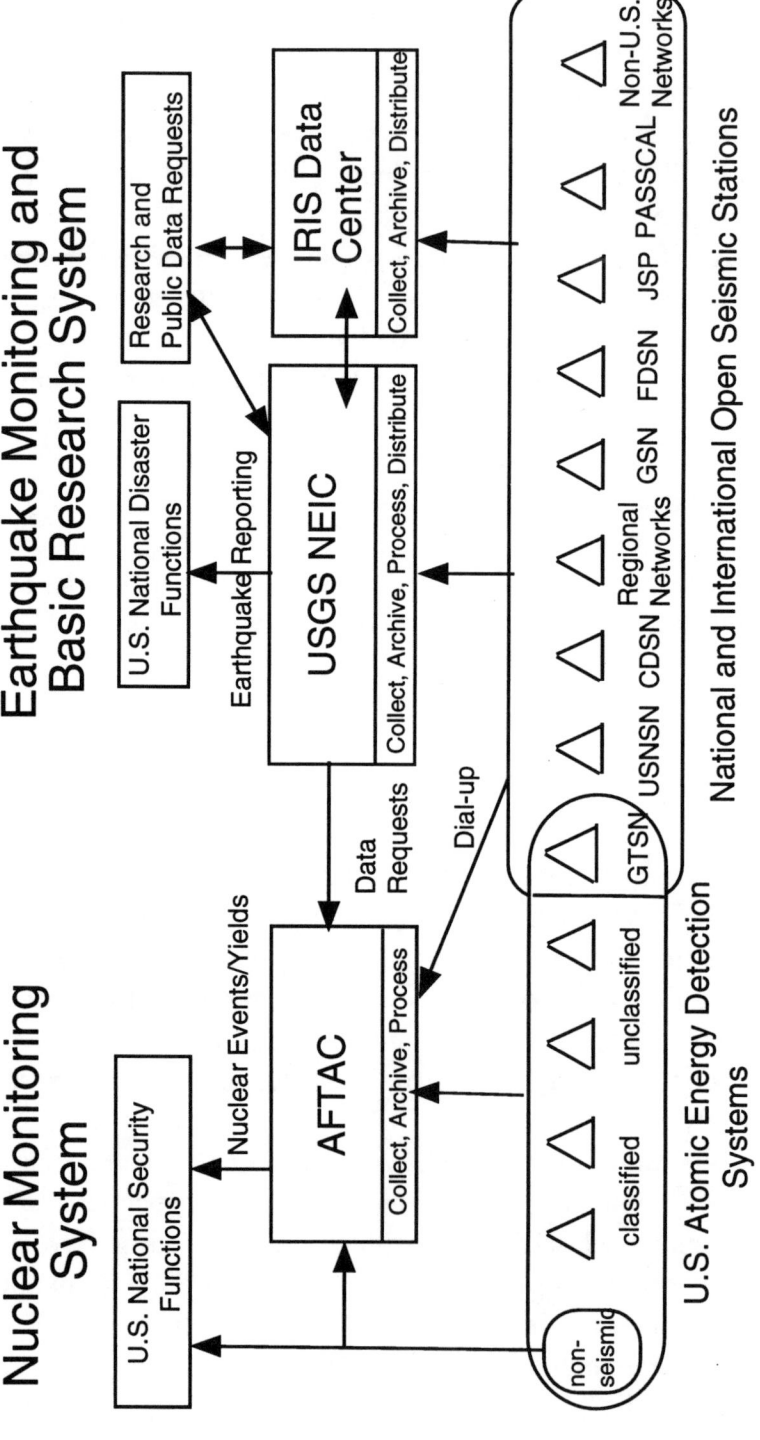

Figure 2.2. Highly schematic versions of the current seismological systems for U.S. nuclear test treaty monitoring, earthquake monitoring, and basic research. At present, very little of the data used in the nuclear test monitoring arena is available to the earthquake monitoring and basic research communities. Strong data exchange and resource coordination exists between the earthquake monitoring and basic research efforts. Some of the complexities that are not shown involve data centers at various universities that collect, archive, process, and distribute data. These coordinate with the USGS earthquake monitoring efforts (see Figure 2.4). Tsunami warning centers in Alaska and Hawaii also access data from the open systems and perform rapid analysis of the signals to assess tsunami potential of large earthquakes. Other nations also operate seismographic networks and data processing centers. Acronyms are defined in Appendix C.

# INTRODUCTION

support earthquake monitoring, but meeting the demands of CTBT monitoring requires an external research and development program that can contribute to and profit from the efforts of the broader research community.

Figure 2.3 shows the planned distribution of the primary stations for the GSETT-3 operation, with those stations actually providing data to the ISMS-IDC as of March 1, 1995, being highlighted. Seventeen of the stations have only three-component broadband instruments, and 15 include a broadband instrument and an array of short-period vertical sensors. The global distribution of stations is expected to improve continually and to number about 50 primary stations or arrays (49 were committed at the time this report was prepared). As many as 100 auxiliary stations are planned as well, with some of these being drawn from the existing global distribution of broadband stations of the Federation of Digital Seismological Networks discussed below. At present, about 40 such stations around the world have dial-up access capability. The value of the auxiliary stations is often assessed in terms of enhanced location capabilities of the ISMS; but their principal value may well lie in the additional identification capabilities that they provide to the U.S. national verification function.

The U.S. earthquake monitoring system is a distributed operation involving many organizations (greatly simplified in Figure 2.2). This effort is supported and operated primarily by the USGS and the NSF-funded IRIS, in collaboration with many university and private-sector efforts. Other government programs involved in earthquake monitoring include the National Oceanographic and Atmospheric Administration (NOAA), the Nuclear Regulatory Commission (USNRC), Federal Emergency Management Agency (FEMA), and many state agencies. The National Seismic System (Heaton et al. 1989) involves coordination of the large number (> 1000) of regional network stations in the United States (Figure 2.4) operated by the USGS and several collaborating universities. The USGS also operates the National Seismic Network (NSN), which is a growing network that will involve about 50 broadband stations deployed within North America. These USGS seismic stations are primarily intended for earthquake monitoring in the seismogenic zones of the country, but the improving accessibility of data from these operations has enabled important basic research applications on global earth structure and earthquake source processes.

This U.S. effort is, in turn, part of a larger international effort that has many organizations and collaborative arrangements. Numerous international and national seismographic networks are involved, ranging from isolated stations to dense regional networks of short-period seismometers to sparse global networks of broadband seismometers. Thousands of seismic stations contribute data to the global system, as illustrated by a map of stations contributing data to the International Seismic Centre (ISC) in Figure 2.5. Several data centers acquire, archive, process, and distribute seismic

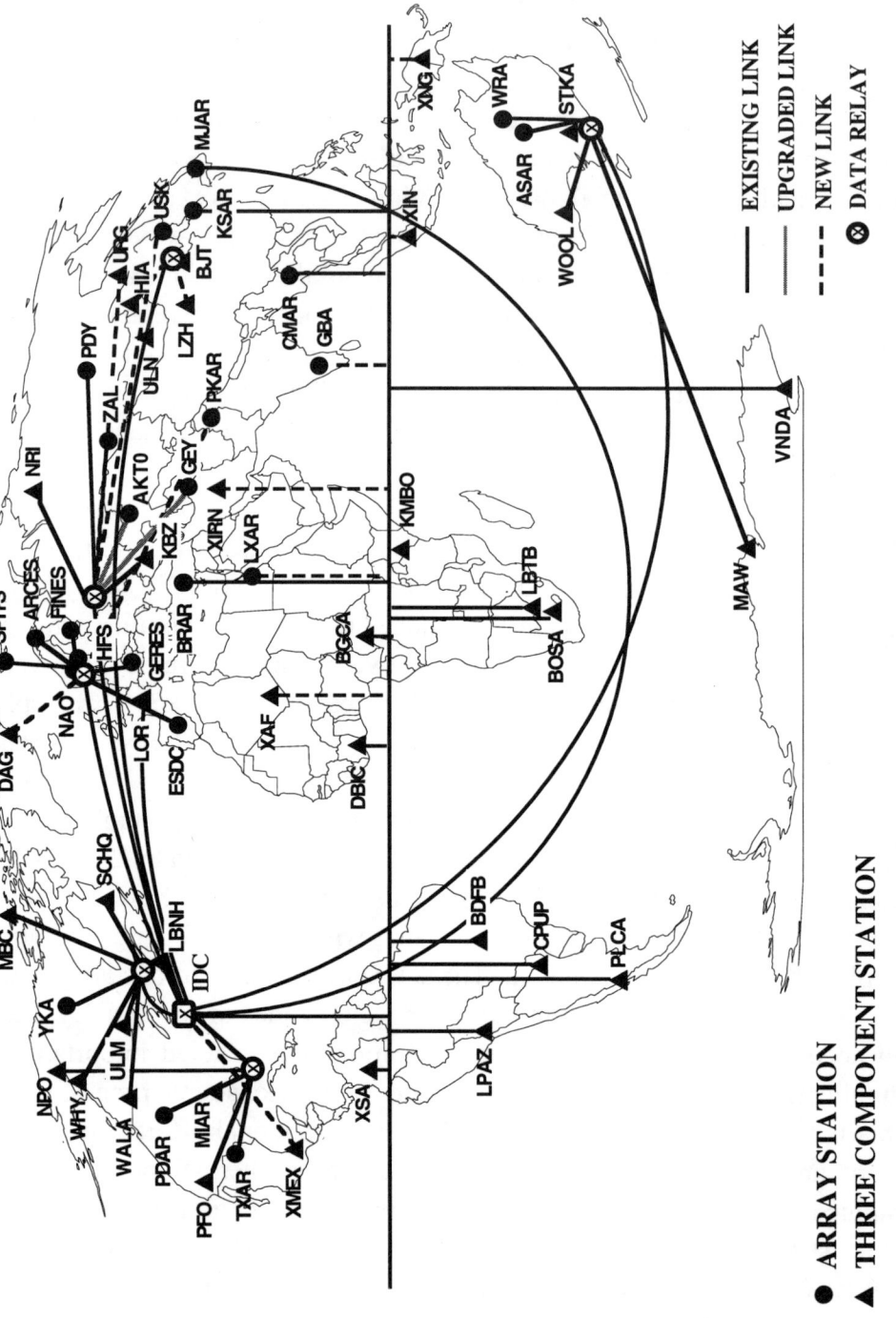

Figure 2.3. Map of the existing and planned primary stations of the GSETT-3 network. Stations that were providing data to the IDC as of 1 March 1995 have their names enclosed by a box (based on Conference on Disarmament CD/1296, 7 March 1995). The data are continuously telemetered to the IDC in Arlington, Virginia, by a variety of telemetry links.

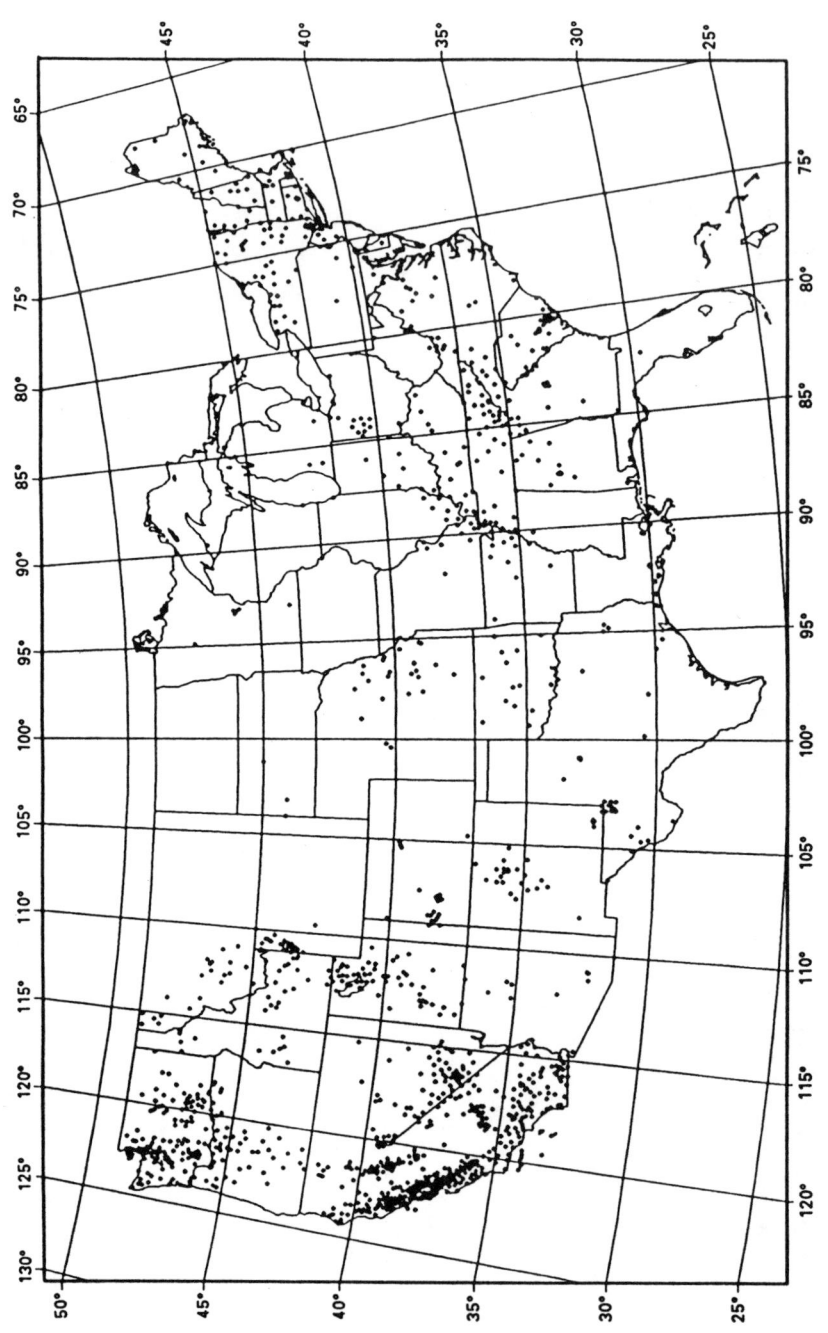

Figure 2.4. A map showing the short-period regional networks operated by the U.S. Geological Survey and various participating universities. These stations are used in routine earthquake monitoring in the seismogenic regions of the United States and typically involve digital recording, continuous telemetry to central processing sites, and automated arrival detection and earthquake location procedures. Basic research efforts using the large number of stations as a large-aperture seismic array have recently produced advances in our understanding of global earth structure and global earthquake ruptures as well. Even more extensive networks of strong-motion instruments exist across the country, although most of their data are not recorded continuously as in the regional networks. Many nations operate short-period regional networks for their own earthquake monitoring applications, and data from these networks may have many CTBT monitoring applications.

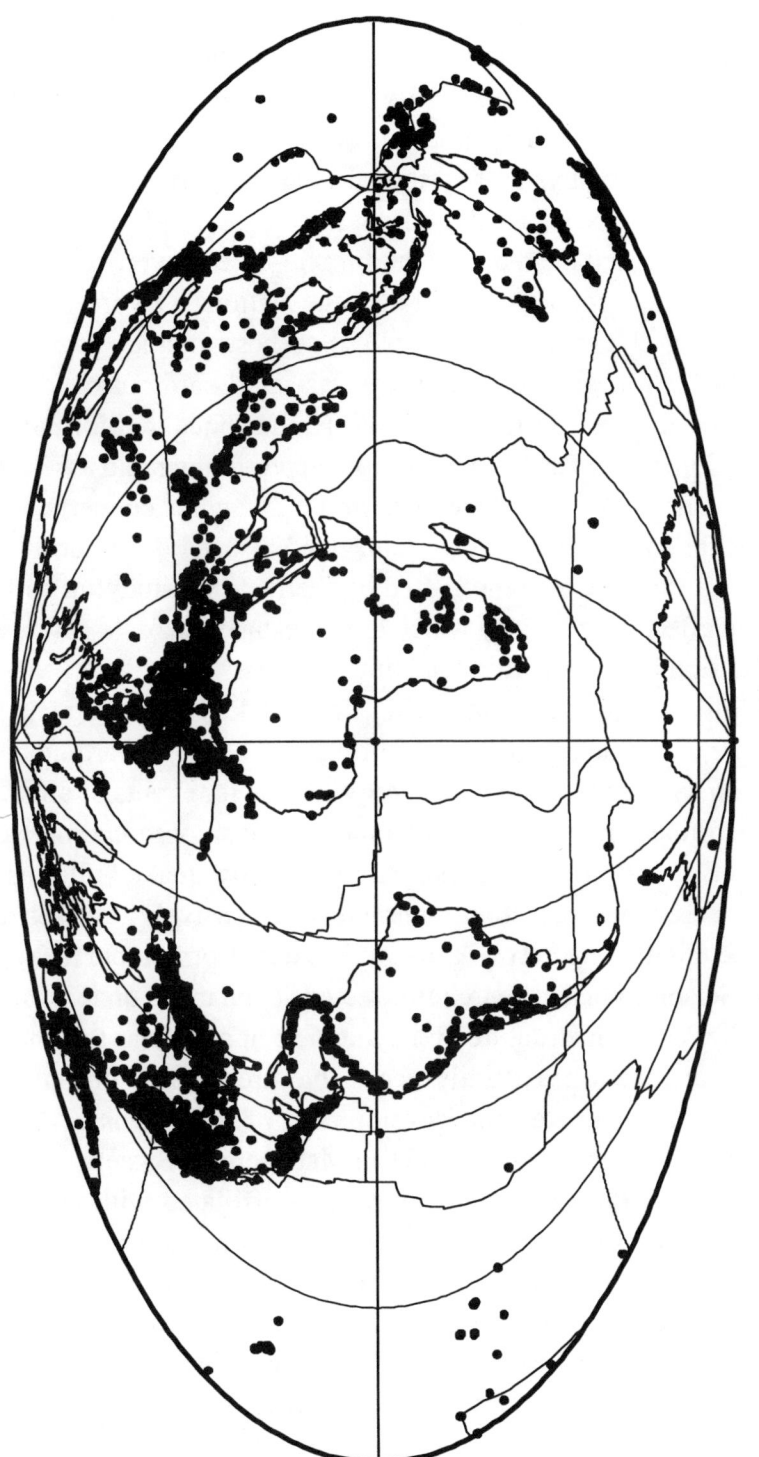

Figure 2.5. A map of the global distribution of stations that reported phase arrival times to the International Seismological Centre (ISC) from 1964 to 1992 (a total of 5,962 stations). In any given year, only about a quarter of this number of stations report, but each year around 200 new stations are added and 200 drop out. After being entered into a standard computer format, the arrival times are used to produce the ISC bulletin of global earthquake locations as well as in inversions for the internal structure of Earth.

data; the size, complexity, communications capabilities, and funding limitations of the system have precluded consolidation of all data into a single earthquake monitoring data center. Nevertheless, a remarkable amount of data and the analyses of that data are shared internationally within days and weeks of the time an event occurs.

While production of definitive bulletins of event parameters on time scales of days to years is one of the primary objectives of the earthquake monitoring system, rapid location and analysis of earthquakes are important for emergency response and hazard mitigation (NRC Real-Time Seismology, 1991; Heaton et al., 1989). These efforts have increased the requirement for real-time data processing of both regional and global seismic data for activities of both the National Oceanographic and Atmospheric Administration (NOAA) and the USGS. These include tsunami warning systems operated in Hawaii and Alaska as well as rapid earthquake location and magnitude estimation performed by the USGS National Earthquake Information Center (NEIC) in Golden, Colorado. USGS has a major role in providing rapid assessment of international earthquake disasters to the State Department, which is concerned about issues such as political stability of the stricken country and disaster assistance. The USGS efforts in global monitoring are also motivated by the fact that studying earthquakes around the world is an effective means by which to understand the basic nature of these phenomena and the natural hazards within the United States.

Unlike nuclear test monitoring, earthquake monitoring and basic seismological research are concerned with precise information about all earthquake activity, including the location, type of faulting, and energy release for events of all sizes. Many countries operate additional regional and international seismographic networks and share data with the NEIC and the ISC for preparation of earthquake bulletins.

The seismic research community extensively utilizes seismic data from earthquake monitoring networks as well as data from global arrays deployed for basic research. Seismological research is directed at enhanced understanding of earthquakes, basic studies of earth structure and dynamics, and nuclear test monitoring. The nature of this research requires readily accessible archives of current and past data. This requirement has prompted the development of the extensive Incorporated Research Institutions for Seismology (IRIS) Data Management System (DMS) as well as several regional network data management systems affiliated with universities. Much of the important international broadband seismic data has been centralized in the past few years. Many international broadband networks are coordinated under the Federation of Digital Seismographic Networks (FDSN), which now involves more than 100 globally distributed state-of-the-art broadband seismic observatories (Figure 2.6). The FDSN data are all archived and distributed by the IRIS-DMS, which effectively serves as the primary global data center for seismological research (Figure 2.2), supplemented by USGS and university data centers. The diverse data requirements of various research

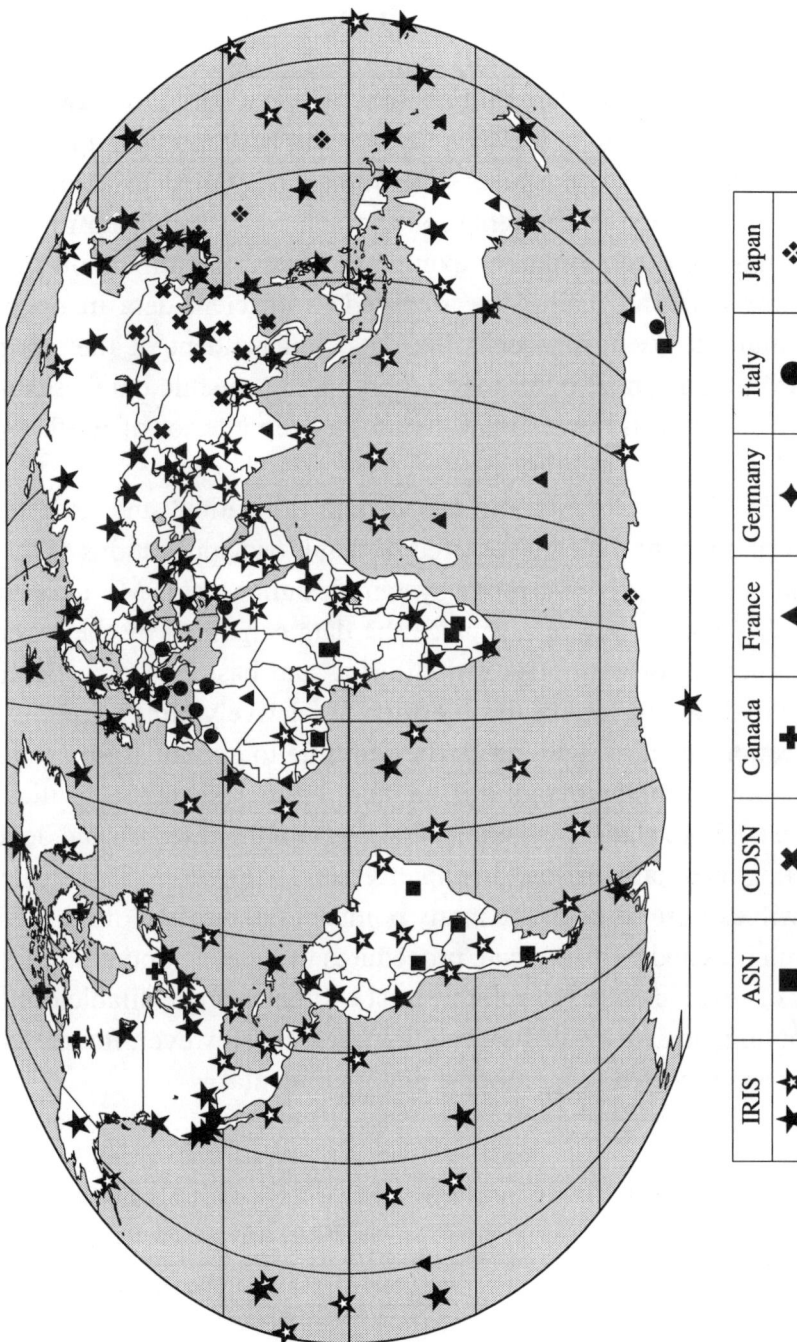

Figure 2.6. A map of the existing and planned (open stars) global distribution of broadband seismographic stations comprising the Federation of Digital Seismographic Networks (FDSN). The U.S. is directly involved in the IRIS, ASN (GTSN), and CDSN networks. The seismic recordings from the FDSN are available through the IRIS Data Management System as well as through data management systems associated with different national efforts. The data from these stations are used for global earthquake monitoring and earthquake rupture analysis, along with numerous applications in the study of shallow and deep earth structure. A limited number of stations have continuous telemetry or telephone dial-up of the signals (around 35 stations of the IRIS network are accessible for rapid analysis of earthquake signals at this time). The high-quality data from these stations have many applications for CTBT monitoring, including regional earthquake analysis, calibration of propagation paths in the areas of monitoring interest, and the recordings that will exist for events for which the ISMS system has difficulties in reliable discrimination. (Figure courtesy of Rhett Butler).

applications make it valuable to maintain continuous data archives on-line whenever possible, and this has strongly influenced the strategy of the IRIS-DMS.

Nuclear monitoring, earthquake monitoring, and basic seismological research activities involve different agencies, data collection and analysis philosophies, and levels of funding, yet they share the unifying attribute of having continuous ground motion recordings as their primary data sources. As long as seismic instrumentation incorporates current technological capabilities that achieve large bandwidth and dynamic range in the recording system, the seismic data will have multiple applications. In the past, data collected for one purpose or another have failed to achieve their maximum potential due to limited instrumentation characteristics and/or limited access to the data. There is now no technological excuse for this underutilization of data, because digital seismic data can readily be archived in efficient data management systems that allow multiple users to access the data, independent of their primary objectives. Thus, ISMS data can be combined with existing seismic databases to the benefit of earthquake analysis and investigations of the deep interior of Earth as well as hazard studies and nuclear test monitoring.

Planning, commitment to achieving broad data utilization, and an effective means for widely distributing the data are required so that broad applications of the data are not negated by an unnecessarily restrictive system design. Most ISMS data will be of high quality, but they cannot begin to replace the data generated by the extensive seismological infrastructure for earthquake monitoring and basic research described above. However, ISMS data will benefit those efforts at relatively minor expense. The U.S. national verification function will similarly continue to benefit from reciprocal access to stations from earthquake monitoring and basic research activities (for example, as a backup to ISMS stations when needed), as will the research and development efforts supporting national verification capabilities.

This report will explore some of the many points of intersection of the different seismological communities and will advocate procedures that enable optimal utilization of the various types of seismic data. No reliable cost estimates are available either for handling and distributing the data or for funding the research. However, increases over present expenditures are expected to be modest and incremental.

# 3

# ISMS DATA CHARACTERISTICS

This chapter addresses the first charge to the panel, involving the data characteristics for ISMS monitoring stations as proposed by the Group of Scientific Experts (GSE). After providing introductory background, the chapter presents comments and recommendations on the technical requirements of ISMS stations and follows that with a discussion of the data streams.

The ISMS data will be an important contribution to the total seismic data needed by diverse parts of the research community. Conversely, much data from current networks can provide essential input to the monitoring effort. It is important, therefore, that the data characteristics of the new ISMS stations be as compatible with the broad needs of seismology in general as is consistent with the objectives of CTBT monitoring.

Furthermore, timely general access to the data streams from broadband three-component instruments at the primary and auxiliary stations of the ISMS will allow the best utilization of the data in earthquake monitoring, research, and treaty monitoring activities to benefit the country. Specific suggestions about ensuring rapid and full access will be discussed in Chapter 4.

## Introduction and Background

The panel was asked to address the following charge:

*Data Characteristics*. The Group of Scientific Experts (GSE) has written draft requirements for an ISMS-standard station that specify characteristics such as sample rate, passband, dynamic range, and sensitivity. They have also proposed a primary network configuration and rough requirements for signal detection, parameter extraction, and event location. What types of data (raw and/or processed) are sought by the seismological community for use in test-ban monitoring research and in other types of basic research?

The draft requirements for an ISMS-standard three-component station were extracted from the Progress Report to the Conference on Disarmament CD/1211 and were circulated to the panel with a request for broad technical feedback. Our comments in this chapter are confined to the specifics of instrumentation in the context of multiuse potential of the seismic data streams. We do not address the CD strategy of a tiered seismic network of primary and auxiliary stations, but we do consider the broad implications of the instrument specifications that have been proposed for ISMS standard stations. It is assumed that these technical specifications apply to the primary stations and that the auxiliary station requirements may not be exactly the same. We will note some distinctions in design requirements for broadband stations versus short-period array components, given the different opportunities for noise-suppression processing.

The issue of what types of data are sought by the seismological community is complex because many components of this community have distinct data requirements, and no single network can service all functions. As a consequence, an extensive and multifaceted national and international infrastructure exists for collecting local, regional, and global seismic data with varying technical specifications (see Chapter 2). This infrastructure supports research and monitoring functions associated with earthquake hazard mitigation, earthquake engineering, fundamental earthquake investigations, local and global earth structure investigations, and earthquake and tsunami warning systems, as well as nuclear test monitoring systems. Many federal, state, university, private, and international organizations are involved. Although each seismological application has its own special data requirements, there are significant intersections in requirements, and the U.S. and international seismic communities have extensive multiagency, multiuniversity, bilateral, and multinational agreements in place to take advantage of the entire data acquisition effort. The panel's perspective, therefore, is that it is most efficient and cost effective to optimize instrument and data compatibility to the extent possible to enable multiple uses of the seismic data.

The ISMS operation will provide both improved real-time access to some existing international seismic data and access to totally new seismic data, which can potentially complement the existing data used in diverse seismic research and monitoring activities. From the research community's point of view, it is critical that international planning for ISMS stations be well coordinated with the existing seismological infrastructure servicing other areas of national need. Feedback to the panel indicates that the broad seismological community agrees that ISMS primary station data can contribute to many of these areas outside the treaty monitoring arena, as long as suitable access is provided.

The panel's response to the first charge addresses proposed attributes of the technical instrumentation of ISMS standard stations, indicating ways in which minor modifications to the technical requirements will optimize multiple-use applications of the seismic data streams. Then, we address the prioritization of the data generated by the ISMS system for both research and earthquake monitoring applications.

## Discussion of ISMS Station Technical Requirements

ISMS instrument specifications emphasize the short-period end of the seismological spectrum, essential for recording small events, whereas the research community emphasizes recording a very broad spectrum of seismic signals with a dynamic range sufficient to resolve ground noise and to record, on scale, the largest signals. The panel has not addressed the seismological objectives that drive the technical requirements for the ISMS stations, but they generally appear to be consistent with the needs of a treaty monitoring system. Fortunately, modern seismic instrumentation has expanded the bandwidth and dynamic range of commonly available sensors. ISMS technical specifications for the three-component broadband sensors to be deployed at each primary station are fairly close to those of the instrumentation deployed by the U.S. Geological Survey (USGS), Incorporated Research Institutions for Seismology (IRIS), and other members of the Federation of Digital Seismographic Networks (FDSN) for a broad range of seismological applications.

The panel strongly endorses the planned inclusion of at least one three-component broadband set of instruments in a low-noise environment at each primary station. The resulting continuous three-component broadband data from the ISMS would have maximum impact in basic research investigations. The broadband channels, with the specified passband, can readily be incorporated into the USGS's earthquake monitoring and analysis procedures, extending the data available for use in near real-time. However, the "Station Requirements for an ISMS Standard Station," listed in Table 1 of CD/1211, are somewhat restrictive and limit broader applications of the data stream. Below, we discuss the relationship of the technical requirements listed in CD/1211 and indicate their relationship to the types of seismic data sought by the seismological community. We note that the GSE has actively been considering some of the recommendations made below (based on the panel's first preliminary report), and new station requirements are proposed in more recent CD working documents, some of which accommodate our suggested changes.

a. *Passband (0.02 to 20 Hz)*. The low-frequency cutoff of this passband will preclude recording of very-long-period surface waves and free oscillations, which are of extensive use in basic research on earthquake sources and earth structure. It is now technologically straightforward to extend the low-frequency response to 0.003 Hz (the lowest frequency of seismological interest) without significant impact on the cost of the broadband instrumentation. This change would replicate the low-frequency response capabilities of many of the IRIS, USGS, and FDSN stations. Since many of the stations initially designated as primary or auxiliary stations for the GSETT-3 already have very broadband sensors (STS-1 or KS54000) that intrinsically achieve a low-frequency response, down to 0.003 Hz, it is straightforward to broaden the specified bandwidth, but it is not possible at present to meet the full desired bandwidth using a single instrument. The panel recognizes that it may be desirable to use only a single sensor when possible. At the very least, instruments such as the STS-2 should be utilized to extend the low frequency response to 0.01 Hz while still having adequate response in the short-period range. ISMS installations with both short-period arrays and a single broadband sensor should have some flexibility in the response criteria of the broadband system to ensure that low-frequency response is not sacrificed unnecessarily. The panel was very concerned to see that some of the updated working documents of the GSE specified low-frequency response down to only 0.04 Hz. This would greatly diminish the broader applications of the data and would even jeopardize the application of the ISMS data for routine functions such as computation of Ms, the surface wave magnitude, which is important for event discrimination.

***The panel recommends that the low-frequency end of the passband of ISMS broadband systems extend to 0.003 Hz wherever possible.***

There are some current applications for the seismic energy above 5 Hz in the earthquake monitoring and basic research communities, and the availability of globally distributed high-frequency data from quiet sites offers new potential for research on earthquakes and earth structure. Many current stations of the IRIS, USGS, and FDSN networks readily can be (or have been) modified to achieve the high-frequency bandwidth of the ISMS specifications at relatively minor cost, and they will complement the primary stations.

b. *Seismometer Noise (10 dB Below Peterson's Low Earth Noise Model (LNM))*. If this criterion is intended for the entire passband (0.02 to 20 Hz), it requires new instrument development, as we know of no broadband seismometer with a noise figure that is 10 dB below the LNM over the full range. A combination of STS-1 or STS-2 and GS-13 instrumentation can achieve this requirement over the passband 0.01 to 10 Hz, and a combined KS54000I and GS-13 can achieve this requirement over the passband 0.03 to 10 Hz. Certainly, this requirement, if achieved, is compatible with all applications of the seismic data.

***It would not seriously impact the research community's current use of the data if the requirement were relaxed in the 10 to 20 Hz range to allow use of existing state-of-the-art equipment in the ISMS, and the panel so recommends.***

# ISMS DATA CHARACTERISTICS

c. *Calibration (within 5 percent in amplitude and 5 degrees in phase)*. This is satisfactory for most uses of the ISMS data. It is satisfied by current IRIS, USGS, and FDSN stations.

***No change is recommended.***

d. *Sample Rate (40 sps + 50 ms)*. This sample rate is suitable for most multiuse applications of ISMS data, as is the lower sampling rate of 20 sps as proposed in more recent GSE documents. The panel notes that in practice the 20-Hz high-frequency response objective is incompatible with the specified sample rate (40 sps). It may be desirable to oversample and filter to remove 50 to 60 Hz noise, and then resample to obtain the final desired high-frequency response. This could reduce the spectral density of the noise, which would benefit all applications of the short-period energy. Although the panel has not chosen to address the monitoring motivation behind the specification, some members noted that the preferred sampling rate is too low for research on and application of spectral characteristics to assist in discriminating between mining blasts and single explosions. Sampling intervals as much as four times smaller than the time between individual blasts in the mining explosions are needed, indicating sampling at up to 100 sps. It is not currently realistic to achieve this sampling rate for continuous ISMS data at all stations; however, it may be desirable for certain stations, especially near mining areas. This high-sample-rate data could be saved on-site and accessed on demand.

***The panel recommends that the sample rate be reevaluated.***

e. *Resolution (18 dB below Peterson's LNM)*. This requirement is driven by the desire to resolve very low amplitude signals that can be enhanced via stacking array elements. While this resolution can be achieved at existing FDSN stations using combined very broadband (VBB) and very short period (VSP) sensors, for isolated three-component broadband stations this level of resolution significantly departs from the needs of present seismological applications. The panel sees no advantage to digitizing deeply into the noise for single broadband stations. Furthermore, there are significant negative consequences. Large earthquakes at teleseismic distances, and even moderate ones at regional distances, produce signals that exceed the finite dynamic range of a 24-bit system, leading to clipping if the system digitizes deeply into the noise.

***The panel recommends that this requirement be relaxed for broadband three-component stations. The panel also recommends that the resolution level be tied to local noise. Since the LNM is rarely achieved, a more expeditious use of the dynamic bandwidth would be to base the floor of the resolution on a site's actual noise levels.***

Alternatively, the suggested CD noise resolution can be achieved by the addition of triggered broadband strong-motion sensors at the broadband stations. This can be implemented at low cost (as is a common practice at IRIS stations in seismically active regions), and it will ensure that ISMS data are not depleted in the very signals of greatest multipurpose use.

f. *Sensitivity (200 counts/nm at 3 Hz)*. There is no known quantizer that can achieve the desired resolution over the entire passband with a sensitivity of 200 counts/nm at 3 Hz. (The sensitivity would have to be set to 800 counts/nm to meet the resolution requirement at 20 Hz.) Sensitivity is adjustable depending on site noise, but for nuclear test discrimination at regional distances the critical passband appears to be in the range of 5 to 8 Hz.

***The panel recommends specifying sensitivity goals at slightly higher frequencies or over a range of frequencies.***

g. *System Noise (10 dB below Peterson's LNM)*. The seismometer noise and sensitivity setting determines the ability to meet this requirement.

***While arrays can take advantage of low system noise to beat down natural background noise, this is not viable for individual three-component systems, so this requirement could be relaxed for the latter.***

A more relevant reference point for system noise requirements is the local site noise, not the LNM.

***The panel recommends that the system noise requirement should clearly specify the frequency band of importance.***

h. *Dynamic Range (126 dB)*. The panel interprets this design goal to apply to the digitizer capability. Existing widely used 24-bit quantizers achieve this range, so this requirement is compatible with multiuse applications of the data stream. However, if dynamic range is defined as the range from the LNM to the clip level, for the desired sensitivity (200 counts/nm at 3 Hz) this system will achieve no more than 111 dB at 20 Hz and 96 dB at 1 Hz. More recent GSE documents propose a more realistic 96 dB requirement.

***No specific recommendation at this time.***

i. *Linearity (90 dB over the passband)*. This is fully compatible with general seismic data requirements for diverse applications of the data.

***No change recommended.***

# ISMS DATA CHARACTERISTICS

j. *Timing Accuracy (1 msec)*. This is fully compatible with general seismic data requirements for diverse applications of the data.
***No change recommended.***

k. *Operating Temperature (-10° to 45°C)*. A low-temperature requirement is probably needed only for certain sites and can be attained for most broadband systems with special insulating techniques and temperature controllers. Provided appropriate power is available, the proper environmental control system can increase the range of station locations, as is desirable for multiuse applications of the seismic data.
***No specific recommendation at this time.***

l. *Authentication (required)*. This is generally not needed for research applications. To the extent consistent with the monitoring goals, measures that are implemented should be such that they do not affect general use of the data stream. Some authentication procedures could involve significant modifications of existing instrumentation. The associated costs may limit the number of stations participating in the ISMS, which has negative implications for system performance.
***No specific recommendation at this time.***

m. *State of Health (at least clock status, calibration status, and vault status)*. Such information is routine and desirable.
***No change recommended.***

n. *Format (one of the formats of the Group of Scientific Experts)*. The broad international seismological community has established a standardized digital seismic data-exchange format, Standard for the Exchange of Earthquake Data (SEED), which is now widely used in the FDSN.
***The panel recommends that mechanism be established that would provide ISMS data in SEED format in addition to other formats that might be used.***

o. *Protocol (Telecommunication Protocol/Information Protocol (TCP/IP)*. This is compatible with other systems.
***No change recommended.***

p. *Delay in Transmission (<15 sec)*. This is compatible with the needs of other systems. For global tsunami warning and earthquake hazard assessment, access within a few minutes is desirable. However, regional earthquake monitoring benefits from delay times of no more than a few seconds.
***No change recommended.***

q. *Data Frame Length (<1 sec)*. This is shorter than in many existing stations, and it is not clear that such frames are an advantage. It is not needed for other uses of seismic data, which typically have data frame lengths of 2.15 to 8.6 sec for 20 sps and 1.075 to 4.3 sec for 40 sps. Shorter frame lengths would negatively impact the data compression schemes used in many existing stations. More recent GSE documents have relaxed this requirement to < 60 sec.
**The panel recommends reconsideration of data frame length requirement.**

r. *Data Access (Priority to International Data Center (IDC), then National Data Center (NDC))*. This item pertains to priority for communication with the station, not end-use distribution of the data. Provision of this access is technologically straightforward and can be implemented on existing systems such as those of the FDSN. It is very desirable that all data recorded at ISMS stations be made available promptly to the general research community.
**No specific recommendation.**

s. *Disk Buffer (7 days)*. This is readily achieved with current technology.
**No change recommended.**

t. *Data Availability (>99 percent)*. This high percentage of reliability is driven by the needs of the nuclear monitoring function. This requirement diverges in practice from many other data acquisition systems because it is not cost effective. The panel expects that a 99 percent data availability requirement will lead to high operation and maintenance costs, thus limiting the funds available to support a large number of stations in the monitoring system, particularly in the auxiliary network. Even the 95% availability recommended for the auxiliary network in some recent GSE documents is likely to prove unduly restrictive. For example, the overall average IRIS network data availability is approximately 90 percent. The stated requirement could preclude use of these high-quality stations as part of the ISMS unless new funds are provided for the necessary level of maintenance. Academic and earthquake monitoring efforts typically prefer data from relatively dense and widely distributed networks, tolerating delayed access to some data and some gaps from individual stations, rather than the ISMS concept of data from a sparse network of highly reliable stations in near real time. With respect to auxiliary stations, tolerating reduced data availability, say at the 90 percent level, from an enhanced number of stations would provide data that would still achieve the overall desired availability and that would be of greater use to the seismological research community because of the expanded coverage. For example, given that the signal-to-noise ratio is such that two stations with 90% reliability have recordable ground motion, the probability that at least one of them will actually record is 99%.

ISMS DATA CHARACTERISTICS 31

Verification researchers also would benefit from the denser coverage and improved understanding of the regional geology and the wave propagation characteristics in interpreting data from an extended seismic network.

***The panel recommends that separate and realistic data availability requirements be established for the primary and auxiliary networks.***

u.  *Timely Data Transmission (>98 percent)*. The availability of real-time data will not only benefit ISMS operations but will also extend the real-time data available to the USGS earthquake monitoring and basic research communities. Experience with four Global Test Seismic Network (GTSN) stations indicates a long-haul communications link availability of 75 to 97 percent, depending on the station, so implementing this level of performance of real-time transmission for all primary stations will be very challenging.

***No specific recommendation at this time.***

v.  *Station Location (within 100 m, array elements within 1 m relative)*. This is a routine requirement, although a specific reference frame for location, precision, and accuracy should be given.

***No specific recommendation.***

w.  *Seismometer Orientation (known within 1 degree)*. This is an extraordinarily high accuracy, not routinely achieved with any borehole instrument (KS36000I and 54000I orientation is +/- 3 degrees), and vault-type instruments can be oriented this accurately only if a suitably accurate survey mark is provided in the vault. More recent GSE documents suggest that 3° is an acceptable specification.

***The panel recommends relaxation of the orientation tolerance.***

## Desired Raw and Processed ISMS Data Streams

This issue will be taken up in detail in the next chapter, but some initial response is warranted in the context of the first charge. There is wide enthusiasm for timely and straightforward access to the broadband three-component data streams from the (continuous) primary and (segmented) auxiliary stations of the ISMS, as well as to continuous data from the auxiliary stations, which will not be collected by the ISMS. The continuous primary data can be directly incorporated into real-time analyses conducted by the USGS global earthquake monitoring system, complementing data collected through other networks. In addition, the broad seismological research community has numerous applications of real-time seismic data analysis, and timely access to the ISMS data would enable maximum utilization of the data in diverse applications. Segmented auxiliary station data, which will be acquired within several hours of events of interest, also will be valuable for rapid analysis by the USGS and basic research communities. Specific suggestions as to how to ensure rapid and full access to the complete waveforms from the broadband sensors in the primary array, as well as all data from the auxiliary network, will be provided in the next chapter.

The ISMS processing will include automated and analyst-reviewed measurements of phase arrival times, slowness measurement from the primary arrays, and array beams, f-k spectra, and event locations. The USGS indicates that the phase and $dt/d\Delta$ measurements and bulletins from the IDC, both automatic and reviewed, would be useful, but not essential, resources for their operational program. The USGS also indicates it is unlikely that it will have serious use for products such as f-k spectra in the near term. USGS operations do not currently place a high emphasis on global array data, either raw data or formed beams, but this situation could change in the future.

There is interest from the general research community in obtaining short-period data from the primary arrays. The primary interest is in the original array data for events of $mb > 4$ for the several-minute time window encompassing teleseismic phases. Access to data from the individual array elements, rather than stacked signals, is likely to be of interest to members of the research community, particularly with respect to discrimination and deep earth structure research. However, the vast quantity of data involved is such that it would be costly to duplicate the archive of the full data set. Developing a procedure for accessing the ISMS data archive for tailored user requests of array data seems to be the most attractive option. There does not appear to be a general requirement in the research community for intermediate products such as the 100 continuous beams formed by each array or f-k spectra as long as access to the raw array data is established by some convenient procedure.

This chapter has considered data characteristics proposed by the GSE for the ISMS seismic stations in light of research requirements both for general seismology and for earthquake and nuclear test monitoring and detection. The general objective of the panel's recommendations is to ensure that the ISMS seismic data are as inclusive and as broadly applicable as possible. The panel has therefore recommended some changes in low-noise level requirements, sample rate requirements, sensitivity goals at higher frequencies, and data frame length. The panel also recommends adopting methods for ensuring rapid and full access to data streams. These are discussed at greater length in the next chapter.

This chapter has dealt with the data characteristics of the proposed permanent ISMS monitoring stations. Characteristics of portable instruments have not been covered, but the panel notes that use of appropriate portable apparatus would increase coverage temporarily in an area of particular interest.

# 4

# DISTRIBUTION OF ISMS DATA WITHIN THE UNITED STATES

In this chapter, we address the second charge to the panel, which involves access to ISMS seismological data by the U.S scientific community. The CTBT monitoring system will collect several other types of data (e.g. infrasound, hydroacoustic, chemical) that are not considered here. We consider the mechanisms and infrastructure required for providing broad access to the ISMS seismic data for multiple-use applications, including nuclear test monitoring, earthquake monitoring, and research efforts that support these monitoring functions.

The design of the CTBT monitoring system has significant implications for the future of nuclear monitoring in the United States and for the structure of seismic monitoring and research on a wide variety of topics of vital interest to the United States. Substantial money is about to be spent on the CTBT monitoring infrastructure, and it is desirable that it be spent wisely and effectively to ensure cost-effective usage of the data for a range of applications. To enable multiple uses of the seismic data, it is important to establish convenient pathways for data access that do not interfere with the primary operations of the nuclear test-ban monitoring effort. This report proposes cost-effective strategies that will provide these pathways. The panel's approach designs the U.S. nuclear monitoring effort to take advantage of existing data archival and distribution capabilities that service seismological applications benefiting the nation.

The panel strongly recommends that the U.S. ISMS-NDC coordinate its efforts with the earthquake monitoring operations of the USGS, the data distribution capabilities of IRIS, and the research and development efforts related to treaty monitoring. This concept of an ISMS-NDC with a multi-element data distribution process, rather than an isolated center servicing all functions, can achieve significant cost reductions and will ensure that full access to the ISMS data is sustained.

## Introduction and Background

The panel was asked to consider the following charge:

*Data Access.* The GSE has specified that all authorized users (most likely the ISMS National Data Center in each participating country) have prompt electronic access (perhaps through the ISMS International Data Center) to all raw and processed data. What kind of access would best satisfy the requirements of other operational groups (e.g., for earthquake hazards and tsunami warning)? How should the data be organized (e.g., by region, station, time period; continuous vs. event segments) and made available (e.g., access time scales—minutes or days; and media—electronic or optical)?

The text of this charge was circulated to the panel's liaison representatives and to numerous members of the seismological research community, with a request for feedback on data access issues. The responses underscored the central importance of data access to all members of the earthquake monitoring and research communities and emphasized the need to incorporate planning for broad data access to near real-time and archived data in the design of the ISMS and the U.S. ISMS-NDC. Strong sentiments were expressed that the seismic data of the ISMS, all of which are unclassified, should be available to both the broader research and the earthquake monitoring communities in a timely manner. The seismic data used for past and present nuclear monitoring purposes, many of which are now unclassified, are not accessible for scientific research and thus fail to achieve their maximum impact in both the nuclear monitoring and earthquake monitoring communities. The panel believes that the new context of unclassified data collection for the ISMS provides an opportunity to implement greater usage of the data streams acquired for monitoring nuclear test treaties than has been the case in the past. The recommendations made below optimize multiple use of the ISMS data while furthering the primary mission of monitoring a CTBT.

### Agencies with an Interest in Seismic Data

Because many of the recommendations in this report deal with the handling and ultimate use of data, it is relevant to review the various agencies with intersecting missions and uses of seismic data. Within the federal government there are several agencies involved with seismic monitoring of earthquakes and/or nuclear testing, and several agencies involved in seismic operations and research. The DOD has traditionally held major roles in research and development in support of nuclear monitoring, mainly organized under the Air Force Office of Scientific Research (AFOSR), the Air Force

Phillips Laboratory (AFPL), and the Advanced Research Projects Agency (ARPA). Actual monitoring and national verification operations, along with advanced development research, have been primarily conducted through the Air Force Technical Applications Center (AFTAC). The Department of Energy (DOE) has a long history of source-mechanism, regional propagation, and seismic discrimination research using data from nuclear explosions at the Nevada Test Site and western U.S. earthquakes. Beginning in late 1994, the primary mission for research and development in support of nuclear monitoring was transferred to DOE, with the ARPA seismic monitoring research and development effort scheduled to phase out over the next two years. (AFOSR and AFPL research programs will continue and will, in combination with DOE, provide necessary scientific and technological support to AFTAC; ARPA will continue to support development of the ISMS-IDC). This structure is considered in greater detail in Chapter 5.

Global and national earthquake monitoring and basic research have been supported by the USGS. It has also coordinated on data acquisition with AFTAC. Other federal organizations involved in earthquake monitoring include the National Oceanographic and Atmospheric Administration (NOAA), the Federal Emergency Management Agency (FEMA), the National Institute for Standards and Technology (NIST), and the Nuclear Regulatory Commission (USNRC). The National Science Foundation (NSF) also supports basic seismological research on earthquakes and earth structure. It funds the Incorporated Research Institutions for Seismology (IRIS), which collects, archives, and distributes seismic data from a global array of permanent stations, temporary regional networks, and portable instrumentation. IRIS has also received funding through the DOD explicitly for data acquisition and research related to nuclear monitoring efforts.

The recommendations that are made in the next section address both the seismological waveform information collected by the ISMS and the parametric measurements, such as arrival times and associated event bulletins. We briefly consider the nature of these forms of seismological information to provide a context for the recommendations.

**Seismic Waveform Data**

All sources of rapid change of strain energy in Earth produce seismic waves that propagate throughout the planet. A high-quality recording of ground motion can capture information about natural phenomena such as earthquake faulting, tidal motions, volcanic eruptions, and large landslides, as well as capturing human-induced vibrations, such as those from nuclear and chemical explosions. Thus, seismograms (recordings at fixed locations of ground vibrations as a function of time) provide the basic information required for nuclear test and earthquake monitoring, disaster response to earthquakes, tsunamis, and volcanic eruptions, natural resource development, and basic research into Earth's structure and tectonics processes.

Seismic waveforms are generally quite complex, and the full information content of the signal cannot be reduced to simple parametric measurements, such as the arrival times and amplitudes of discrete phases. The overall wave shape contains valuable information about the source process that generated the disturbance and the interaction of the radiated wavefield with Earth's structure. By designing ground motion sensors that record a wide range of frequencies (broadband sensors), seismologists increase information content in the recorded signals. For about a century, seismologists have been developing source and wave propagation theory and analytic methods to extract information from the completed seismogram. This is now an advanced quantitative science, and, with the latest generation of seismic instrumentation providing nearly complete recording of all ground motions at a given site, every broadband seismogram can provide extensive information about source, path, and receiver- site effects.

For example, seismic waveforms can be used to locate events (using the characteristic sequence and amplitudes of arrivals recorded at varying distances from a source), to determine the orientation of an earthquake fault and the sense of shearing motion during the rupture, to image the variable slip on the fault surface, and to quantify the total energy release during the event. Modeling broadband seismograms identifies characteristics of earth structure, like the heterogeneity of the crust, that influence the signals from a surreptitious nuclear test. For seismograms recorded at regional distances of from 100–1,200 km, which involve complex reverberations in the shallow crust, the relative amplitudes of different arrivals as a function of frequency are among the strongest diagnostics of the source type (e.g., OTA, 1988). These regional signals are of great importance for monitoring a CTBT, because for the smallest events of concern they may be the only data available.

Automation is essential for U.S. CTBT monitoring, as there will be thousands of events that must be detected, located, and identified each year. The advanced state of waveform analysis and processing is such that the longest delay in determining the faulting orientation of large earthquakes around the world tends to be the propagation delay, that is, the time it takes for the seismic waves to arrive at a sufficient number of stations to perform a stable analysis. Rapid access to seismograms from many stations makes it possible to automate many aspects of routine nuclear test and earthquake monitoring.

Most research applications do not require real-time access to waveform data, but they do require the ability to retrieve diverse segments of past recordings. The waveforms are needed because no simple standardized set of archival parameters will service all applications. In some cases, only a few seconds of a given recording may be analyzed, while in other cases many hours of the same record may be used in studying either the source or the earth structure. This has prompted the development of the IRIS-DMS archive of continuous broadband seismic data, which has the ability to service individually tailored data requests. With the role that frequency dependence of waveform energy and path effects play in discriminating nuclear explosion signals from quarry blasts and earthquakes, availability of archives of waveforms from previous regional events is also as critical for nuclear monitoring. This is particularly true when monitoring a region with no prior history of large explosions. Thus, archives of seismic waveforms play a critical role in the nuclear monitoring arena as well.

Broadband three-component waveforms form the primary data base now used in basic earthquake source and earth structure investigations. Typically, broadband data are recorded at relatively isolated observatory stations, although arrays of portable and semi- permanent broadband stations have been deployed recently. The more extensive the global coverage provided by broadband recordings, the more detailed the information about sources and deep structure that can be retrieved. For some applications, only a modest number of global stations are needed, possibly providing data in near real-time (this would apply, for example, to tsunami warning systems), but almost all seismic analysis procedures are enhanced by increasing the number of observations, as long as the quality of those observations is high. Much of the enthusiasm in the research community for access to the ISMS broadband data stems from the enhanced research potential provided by the increased numbers of broadband recordings that will be available from stations located around the world for each earthquake or explosion event.

The ISMS also includes a number of sites with arrays of closely spaced, high-frequency instruments in addition to a centrally located, three-component broadband instrument. By combining the data from the high-frequency instruments in various ways, analysts are able to detect signals from smaller events, identify the type of seismic wave producing the signal, locate the events more accurately, and in some cases, associate overlapping signals from multiple sources with the proper source. The instrumentation at many of these stations is particularly appropriate for the analysis of the higher-frequency wave field generated by small events. Conversely, small arrays are of limited incremental use for studies of earthquake sources and the structure of Earth which require long-period waves.

For effective monitoring of a CTBT, knowledge of crustal structure and earthquake characteristics on a global basis is very important. For example, research on the crustal structure under ISMS stations requires that the research community have access to the data from the ISMS stations. Some of the many ways in which providing broad access to the ISMS data will enable earthquake monitoring and basic research activities that would improve CTBT monitoring are by:

- allowing research on event detection, discrimination, and yield estimation to be conducted using the same data as employed in the operational environment;
- improving our knowledge of earth structure and hence improving the accuracy of event locations, and aiding in regional characterization and in the resolution of ambiguous events;
- augmenting the data available for basic research on regional crustal and upper mantle structure in regions of importance for CTBT monitoring; and
- augmenting the data available for research on earthquake mechanisms, source depths, and scaling properties in regions of importance for CTBT monitoring.

All these activities require the use of waveform data. In addition, the parameter data produced by the operations of the ISMS will be important for preparation of event bulletins, which is discussed next.

**Seismological Event Bulletins**

Both nuclear monitoring and earthquake monitoring involve the preparation of bulletins of events. These bulletins are lists of seismic events, usually arranged chronologically, that give at a minimum the origin time, the event location (latitude, longitude, depth), and one or more seismic magnitudes, all determined from an analysis of the seismic wave arrivals observed at and reported from stations around the world. Some bulletins also include, for each listed event, the arrival times of seismic waves detected at each station. Traditionally, many seismograph stations have reported arrival times regularly to the organizations that publish bulletins.

Various types of global bulletins have been developed over the years. The International Seismological Centre (ISC), located in Newbury, England, publishes the most complete and most accurate bulletin of global seismicity. The ISC bulletin appears about two years in arrears and is based upon information (such as the observed arrival times of various seismic waves) contributed currently from more than 1,800 seismographic stations around the world (see Figure 2.5), most of them recording in analog formats such as ink on paper or photographs. At present, the ISC bulletin is not complete even at magnitude 5.5 in some remote parts of the southern hemisphere. (paper given by Robin Adams, IRIS workshop, 1991).

The USGS National Earthquake Information Center (NEIC) publishes global bulletins known as the Quick Epicenter Determination (QED) and the Preliminary

Determination of Epicenters (PDE). The QED is first produced about an hour in arrears for a very limited number of earthquakes, and more generally about a week in arrears. The PDE appears about a month in arrears. These bulletins are produced on different time scales in order to service different needs of the earthquake monitoring community. The time lag allowed for any bulletin preparation determines the number of data that can be used to detect and locate events, given that many stations are in remote areas with limited communications. For 1991, the NEIC located 16,516 events, including 1,585 events of seismic magnitude 5 and above, and 4,372 events with magnitudes from 4–4.9. That year, the ISC located 1,373 events previously unidentified in bulletins utilizing fewer stations, 34 with magnitudes > 5, 209 with magnitudes 4.4–4.9, 280 with magnitudes 4.0–4.4, and 850 with magnitudes < 4.

The primary stations envisioned for GSETT-3 have been estimated (CD/1254) to have a threshold detection capability in the magnitude range below 3 for parts of Eurasia and North America, above magnitude 3.4 in some continental areas of the southern hemisphere, and above magnitude 3.8 in parts of the southern oceans. From these detections, plus additional data that may be requested from auxiliary stations as deemed necessary to improve location estimates, the ISMS-IDC will obtain automated event locations that will be reviewed and, if necessary, corrected by an analyst. The GSETT-3 IDC is now publishing a bulletin of global seismicity, known as the Reviewed Event Bulletin (REB), two days in arrears. This production schedule is shorter than those of the ISC and the USGS (except for a limited number of earthquakes studied promptly by the USGS, together with IRIS). For treaty-monitoring purposes, it will be important to be able to examine the region of a suspicious seismic event as soon as possible, using available methods, and if suspicions persist, to request and carry out an on-site inspection.

The GSETT-3 REB differs from current USGS and ISC bulletins in ways other than timeliness of production. The REB is based upon the ISMS-IDC's own analysis of digital seismograms communicated in near real time, rather than the ISC's practice of analyzing measurements of wave arrival times made at the contributing stations. The REB uses fewer stations, although some are arrays, to locate events than either the USGS or the ISC. In the early days of GSETT-3, the REB may give event locations that in many cases will not be as accurate as the later-published USGS and ISC bulletins, but as the GSETT network is calibrated, the quality will improve. The REB may have more uniform global coverage than the USGS and ISC, but the latter's bulletins are likely to have improved coverage in certain areas. Certainly it may be expected that in parts of the world where strong national programs exist to study earthquake hazard (e.g., China, Japan, Mexico, and the United States), these latter programs will provide seismicity bulletins far superior to the REB in all attributes except timeliness. As a result, the REB includes events that the USGS and ISC now miss; and conversely, the USGS and the

ISC bulletins include events that are not in the REB. Comparison of the various bulletins should enhance all.

ISMS-NDCs may also contribute supplemental data to the ISMS-IDC. Such data could include seismicity bulletins based upon various regional, national, and international station networks, and the seismic-wave arrival times from analog and digital stations upon which these bulletins are based. For the GSETT-3 experiment, only seismicity bulletins will be provided, and these will be used in assessing the quality and completeness of the REB. However, it does not appear that the ISMS-IDC plans to use these supplemental data in its bulletin preparation because they will usually arrive too late to assist in production of the REB. Furthermore, there does not appear to be a systematic plan to compare the bulletins.

Given the current timing constraints, supplemental data and ISC/USGS operations cannot assist GSETT-3 in production of the REB. However, the data gathered by GSETT-3 could greatly assist the ISC and the USGS in ways that could have a positive impact on the general effort to improve CTBT monitoring. If the REB and its underlying wave picks were made available to the ISC and the USGS, this information could be combined with other data available to those organizations and they could provide more accurate locations than the REB for all seismic events above about magnitude 4.5 (about 2,300 events per year; see Table 4.1) and perhaps almost all events down to magnitude 4 (about 7,100 events per year). As a long-term goal, complete coverage on continents down to magnitude 3 may be achievable, using current and planned stations.

A comprehensive global seismicity bulletin that emphasizes completeness and accuracy of location rather than speed of publication would improve CTBT monitoring by:

- providing location estimates generally superior to those of the REB, thus supporting evaluation of the REB and the preparation of guidance on how to improve it;
- locating events not included in the REB, thus allowing an evaluation of the REB threshold in different regions;
- providing an archive of accurate event locations useful for prompt interpretation of new seismic locations in the REB as they accumulate;
- supporting special studies of seismicity in different regions of Earth that may be of CTBT concern;
- improving our knowledge of earth structure and hence improving the accuracy of event location by the REB; and
- aiding in regional characterization and in the resolution of ambiguous events.

**TABLE 4.1.**

| Seismic magnitude | Number of earthquakes per year (and per day), worldwide, above each magnitude* |
|---|---|
| 4.5 | 2,300 (6) |
| 4.0 | 7,100 (19) |
| 3.5 | 19,000 (52) |
| 3.0 | 68,000 (186) |
| 2.5 | 209,000 (572) |

* Based on Ringdal (1985).

At present, most earthquake bulletins intentionally exclude quarry blast information to avoid contamination of the natural seismicity information. For CTBT monitoring purposes, information about quarry blasts is of great importance, and the REB will include many such sources. It will be a major task of the national verification activity to identify quarry blasts and to ensure that none of the explosions are nuclear tests. The earthquake monitoring community could make a significant contribution to CTBT monitoring by determining quarry blast locations on regional and global scales, perhaps producing a separate bulletin for such events. This would require modification of existing procedures in which many station operators screen out quarry blast information, and there would be a nontrivial cost for the additional operations; however it would provide additional information for identifying the many quarry blasts detected by the ISMS, especially those on U.S. territory.

There are additional general reasons to develop an improved global bulletin using all the data available to the ISC and the USGS-NEIC, augmented by data from the GSETT-3 and the subsequent ISMS. A global bulletin is the primary database summarizing seismic activity for many interested users outside seismology in geophysical research and in quantitative estimation of seismic hazard. (Far more scientists and engineers use seismicity bulletins than use seismograms directly.) For the seismological research community, an improved bulletin with the goal of complete coverage above a certain (low) magnitude threshold would be important for research on earthquake prediction and hazards. It would also focus attention on the need for quiet sites, highly reliable stations, and appropriate station siting—all issues of concern for explosion monitoring.

GSETT-3 is essentially focused around the effort to produce a global bulletin of seismicity only two days in arrears. If the data generated by GSETT-3 for this purpose (in particular the wave picks) were made available to organizations now publishing bulletins weeks and years in arrears, the outcome would likely be significantly improved accuracy of event location and improved global coverage down to lower magnitudes. Since seismicity bulletins are among the most basic and important databases in geophysics and in the study of natural hazard reduction, many scientists and engineers and their clients, including the general public as well as the CTBT monitoring community, would benefit.

## Fundamental Guidelines for Data Access Issues

Effective seismic monitoring of a CTBT requires the detection, location, and identification of underground nuclear explosions with high confidence and a low false-alarm rate. In this context, once an earthquake is clearly identified as such there is no further immediate interest in its signals. (In the longer term, an archive of such signals can assist future discrimination efforts by providing comparison recordings from the same region.) However, these same recordings are of great value for other applications, such as earthquake monitoring, analysis of the earthquake faulting process, and analysis of structure of Earth. If the instrumentation operated by the ISMS includes appropriate dynamic range and bandwidth, the recorded signals are certain to be useful for many purposes in addition to routine screening to detect nuclear explosion signals. (This was the basic premise underlying the recommendations in Chapter 3.)

The ISMS data quality, distribution of stations, digital format, broadband response, large dynamic range, and timely electronic access all contribute to the potential value of the data. In order for the data to fulfill this potential for both the monitoring and the broader seismological communities, they must reside in readily accessible archives. The panel feels strongly that it is in the interest of efficient use of resources to ensure that the ISMS data be accessible to U.S. scientists by means that maximize their usefulness to all efforts of national interest, to the extent possible without compromising the basic mission of the ISMS. The combined user community should attempt to integrate all international seismic data acquisition, archiving, distribution, and bulletin preparation efforts in a way that benefits all of the potential users of seismic data and strengthens monitoring capabilities in both the short and long term. The panel concludes that efficient integration of the ISMS and ISMS-NDC with existing facilities for earthquake monitoring and distribution of data to the seismic research community can provide benefits to the nuclear test and earthquake monitoring communities and to the research efforts that support them.

As the ISMS is developed, it is critical to recognize that there are many existing international arrangements for open access of seismic data, and these should not be undermined by the political sensitivities associated with nuclear monitoring efforts.

***The panel recommends that steps be taken to ensure that the development of the ISMS does not result in a reduction of existing capabilities of the U.S. scientific community.***

The United States should support the policy that incorporation of existing stations into either primary or auxiliary station affiliation with the ISMS should not result in restriction of the current access to these or other stations. This requires that there be either open access to the complete data streams through the ISMS or that arrangements be made for access via previous procedures. Furthermore, the United States should support a policy that establishes similar procedures for new stations. In addition, a nation's participation in the ISMS should not introduce barriers to obtaining data from other stations previously operating in that country. This issue should be addressed in the preparation of the protocol to the treaty.

For the primary stations, access through the ISMS would be advantageous because the rapid, centralized collection of the data will provide signals from a world-wide network of high-quality stations in near real time from a single source. This could benefit disaster mitigation and response efforts as well as facilitate seismic research into a number of areas of real-time seismic data processing, with attendant monitoring, scientific, and societal benefits.

For the auxiliary stations, access to the segmented data from the ISMS will also increase rapid access to many stations, but it is important that alternate means of retrieving the complete continuous data from these stations be maintained or, in the case of new stations, established. This will require coordination between operators of the auxiliary stations and the nuclear monitoring community, which will be investing in these stations, to ensure that on-demand access is available and that the data satisfy the operational requirements for the auxiliary network.

Historically, many of the seismic data collected by the United States for nuclear monitoring have not been accessible to the scientific community, even though the data were unclassified. The ISMS stations and their data will be unclassified even though they will be used for monitoring purposes. It is important that the ISMS data be available to the broader seismic community because this will benefit many activities, including CTBT monitoring. Development of new methods of analysis and testing of nuclear monitoring procedures will be facilitated by access to the actual data that are used in the monitoring operation. This access will enhance the interactions between the nuclear monitoring research and operational communities. This fact has been amply demonstrated in other contexts in the past.

The research community can also play a part in the confidence-building process that is an essential element in the justification of the ISMS. These researchers will be advisors to their governments and will provide important independent checks and balances on the operations of the monitoring system, as well as sources of insight into the geophysical properties of regions of Earth, the nature of specific events of interest, and monitoring methods in general. In addition, the broader the user community is, the better the feedback about quality-control issues and instrumentation problems. Such problems are often revealed in the course of analysis of recordings for large earthquakes, which may be ignored in the national verification effort.

*The panel recommends that the ISMS and nuclear monitoring communities adopt a clear commitment to provide ready access to all seismic data collected by the ISMS, and that language to this effect be inserted into the Protocol to the Treaty.*

The ISMS-IDC will provide to the ISMS-NDCs data that will directly service national nuclear monitoring applications and other activities.

*The panel recommends that all the ISMS data received by the U.S. ISMS-NDC be made available to the earthquake monitoring agencies and the scientific community in a timely manner, as well as to the nuclear monitoring operation.*

To accomplish this, the U.S. ISMS-NDC must be committed to interfacing with both the broader scientific community and the nuclear monitoring community. Language to this effect should be placed in the tasking requirements of the U.S. NDC operating organization or organizations.

*The panel recommends that the U.S. Government establish a multiagency advisory committee, with representation from the earthquake monitoring and basic research communities to facilitate interagency data transmission and to address cost issues.*

The current plan for the ISMS-NDC will maintain an archive of all broadband and array data from the primary and auxiliary stations; however, this will probably not be a readily accessible online archive.

*The panel recommends that the ISMS-NDC forward the data streams that are of greatest interest for other applications from the ISMS-IDC to appropriate earthquake monitoring facilities. The waveform data should be accompanied by associated calibration and station parameter information.*

Reformatting the data to achieve a single archival format is highly desirable.

*The panel recommends that there be no restrictions on the availability of primary and auxiliary data (such as limiting availability to those data provided by the*

***U.S.-operated stations); all ISMS data should be made available. Language to this effect should be included in the CTBT Protocol.***

Although a major interactive data distribution system could be established at the ISMS-NDC to service all user requests for the nuclear monitoring data, this is likely to be very costly and would replicate existing data distribution capabilities. In addition, there is concern that the responsibility for data distribution could be subordinated to the nuclear monitoring operations, making data access difficult in practice. Procedures for minimizing the data distribution burden of the ISMS-NDC are discussed in the next section. The panel has attempted to identify cost-effective pathways that both ensure data access and minimize duplication of effort.

It is also important to consider the facts that the ISMS system will not operate alone and that there will continue to be other seismic data collected that will prove valuable for treaty verification efforts. It is a reality today that many more seismographic stations are operational and capable of contributing significant data on particular seismic events than are included in lists of proposed primary and auxiliary stations. It has been a common practice by our own national verification group at AFTAC to seek additional data beyond the USAEDS networks in the resolution of problem events. The event of interest might have been recorded well by several stations in a regional or national network where only one or two of those stations are designated ISMS stations; hence critically important data may exist beyond those that the IDC will normally access. The open access of all stations, from both the nuclear monitoring and earthquake monitoring arenas, is to be strongly encouraged. In presentations at the Conference on Disarmament in Geneva, the United States has explicitly recognized that "other seismological resources" (CD/NTB/WP.96), beyond those associated with specific national and international monitoring systems, can and should contribute to CTBT verification.

There are therefore accepted operational reasons, as well as reasons stemming from support of research and development activities, for the United States to pursue practical methods of data access to all seismographic stations around the world that meet minimal criteria. The growth of digital seismographic installations in numerous countries, together with expected reductions in the cost of communications in future years, could mean that monitoring of all types of seismic activity will greatly improve in many regions.

The United States can help tap into these resources by supporting the development of simple communications hardware and software and by encouraging a policy of open data access among national and regional seismographic networks in different countries. Seismologists have a tradition of freely exchanging data that goes back decades. Although the IDC will rely upon the ISMS primary and auxiliary networks for routine analysis, access to additional stations on an ad hoc basis is likely to be the key to a better understanding of what would otherwise be problem events.

## Agency-Specific Recommendations Concerning Data Access Issues

We now consider issues associated with each type of data stream arriving at the U.S. ISMS-NDC. This is done in the context of fundamental design of the U.S. ISMS-NDC and of specific agency activities, with the aim of identifying pathways that will implement the generic recommendations of the previous section. The seismological methods for monitoring a CTBT will be developed in three different organizational contexts. These will involve an international process associated with the treaty negotiation itself, a national process, different in each country, that defines the responsibilities of operational agencies (such as AFTAC and the USGS in the United States), and a broader seismological context involving organizations doing seismology that have no formal responsibility to report to the ISMS or to the nuclear monitoring agency. The problem of explosion monitoring and the necessary research and development efforts are very different as seen from these three organizational contexts. The panel has addressed the data access issue from the broader seismological community perspective, emphasizing the way that advances in the general seismological community will have a positive and significant impact on explosion monitoring.

There are two basic strategies that could be pursued for the ISMS-NDC, involving either a stand-alone center that has all data acquisition, processing, and distribution capabilities, or a center that uses existing distributed capabilities to help provide the required functions. For an autonomous center to provide open access to the very large and diverse ISMS data set in its archive, the commitment of significant resources and personnel to service highly variable data requests from the earthquake monitoring and general research communities would be required. The diversity of such data usages should not be underestimated. Even with extensive computer automation and massive data storage capabilities, such a level of activity would have a significant impact on operational activities at the ISMS-NDC. For example, the IRIS-DMS has distributed more than a terabyte of individually tailored data products, spanning the multiyear interval of its on-line data. These resources and procedures could make a major contribution to the distribution of the ISMS data, thereby reducing costs for the ISMS-NDC and facilitating user access.

A concern expressed by many representatives of the earthquake monitoring and general research communities in the United States is that if the treaty monitoring agency provides this service, it will not attach a high priority to providing access to the current and archival data. At the ISMS-NDC, it is likely that the on-line portion of the archive will be limited to the most recent year of records (as will be the case for the GSETT-3), and there may be difficulties in obtaining access to earlier data that has been placed in off-line storage. The concern is that when fiscal limitations are imposed, support for general access to the ISMS data will likely be the first area compromised. Past experience has indicated that operational organizations have to focus on narrow goals, which can poorly serve the broader seismological community (and, in the long run, the U.S. efforts in explosion monitoring). To some extent, this issue can be resolved by explicit tasking of the U.S. ISMS-NDC to ensure that data access is sustained.

Under the structure we envisage for the ISMS-NDC, much of the servicing of data requests could be provided through existing mechanisms, reducing the need to invest in parallel distribution services. In addition, because there are likely to be multiple agencies involved in the data acquisition process, the distributed approach will naturally provide the necessary coordinated effort. The classified national verification functions would be structurally separate from the data distribution process to address data security issues. The panel believes that a self-contained ISMS-NDC would probably be more costly and provide less access to the ISMS data, than an ISMS-NDC with distributed functions requires several levels of interagency coordination. Since the prototype ISMS-NDC explicitly involves interagency coordination between AFTAC and the USGS on data collection, it should be straightforward to coordinate on data distribution and multiple utilization of ISMS data.

The continuous three-component broadband data from the primary stations are certainly of interest to earthquake monitoring groups, such as the USGS and tsunami warning systems; near real-time access to selected stations is necessary for these earthquake monitoring applications.

***The panel recommends that the U.S. ISMS-NDC make continuous primary station broadband data available in near real time to earthquake monitoring agencies; the data should be archived at the IRIS-DMS. The continuous broadband data should be provided with no windowing.***

This approach duplicates the archive of the continuous broadband data (only a small fraction of the total ISMS data), but it will allow data to be stored in a permanent on-line database whose primary mission is to make data available to the broad community. In addition, this approach exploits the vast existing infrastructure of the IRIS-DMS for servicing extensive user requests for broadband data. The IRIS-DMS has a commendable record of promptly supplying data in response to all requests, including those from non-IRIS members. The ISMS-NDC will thereby not need to directly service the user requests for broadband ISMS data, removing a major distribution burden (far greater than the expected demand for array data, at least in the short term) and reducing its data security issues.

Archiving the continuous broadband data at the IRIS-DMS will assemble broadband data from many sources in a single data base and will also allow further standardization of data formats and instrument response information, greatly facilitating user access to the continuous global broadband seismic data. It is critical that data quality-control information from the ISMS accompany the broadband data. Otherwise, redundant and expensive quality control would have to be performed. The costs for the broadband data transmission and archival should be shared by earthquake monitoring agencies and earthquake and nuclear explosion research agencies in proportion to their projected usage.

***The panel recommends that a prototype of the broadband data distribution system be developed in the GSETT-3 experiment.***

At the time this report was written, the USGS had been seeking to obtain the continuous data from primary stations, but was hampered by limited communication links with the ISMS-NDC connection. A direct connection to the prototype ISMS-IDC exists and could be used to pass data to the USGS earthquake monitoring activities until appropriate communications are installed.

The segmented three-component broadband data from the auxiliary stations are also important for earthquake monitoring functions. The ISMS will facilitate rapid access to these data. This is expected to be a relatively small data set that would incur only small incremental costs beyond that for accessing the continuous primary station broadband data. Access to the continuous data at these auxiliary stations, much of which will never be collected by the ISMS but all of which will be of value to the earthquake monitoring and research communities, should continue to be provided to the IRIS-DMC under existing or new arrangements.

***The panel recommends that the U.S. ISMS-NDC make the segmented auxiliary station broadband data that it acquires available in near real time to the earthquake monitoring agencies; the data should be archived at the IRIS-DMS.***

Since many of the ISMS broadband stations are currently operated by members of the FDSN, which has data distribution agreements with IRIS, it should be politically straightforward to incorporate both the continuous primary and segmented auxiliary data from these stations into the IRIS-DMS. For U.S. and non-U.S. ISMS stations that are not associated with IRIS or the FDSN, agreements should be obtained to allow all of their broadband data to be archived by IRIS, as this will help service the archival and data distribution of all high-quality broadband data. It would be desirable to have the continuous broadband data from any auxiliary stations not currently part of the FDSN also provided to the IRIS-DMS.

*The panel recommends that the continuous data from auxiliary stations (most of which will not be accessed routinely by the ISMS) should continue to be archived and distributed through existing procedures of the FDSN. Operational support of the U.S. auxiliary stations should be shared by the nuclear monitoring, earthquake monitoring, and basic research agencies.*

The most technically and financially difficult data access issue involves the short-period primary array data, which have immense storage requirements and pose major challenges for maintaining a fully accessible on-line database. These continuous data are of central importance for nuclear monitoring procedures. The need for short-period array data in the earthquake monitoring and basic research communities is currently rather limited but is almost certain to increase with time. Possible applications of the data include refined earthquake location in certain areas and investigations of deep earth structure using the high-quality array data. It is probable that array data initially will be most desired by the seismological community involved in nuclear test monitoring research, with their access to the data being directly beneficial to CTBT monitoring research. For these scientists, an important issue will be whether the data are archived on-line, on tape, or on other off-line media. The current plan for AFTAC operations during GSETT-3 will be to maintain up to one year of data on line and older data in permanent 8-mm tape storage. There are several approaches to making these data available to the broader community, each with different cost implications.

The first option is to send all of the ISMS data to the USGS or IRIS to distribute through existing procedures. This approach does not involve any direct servicing of user requests by the ISMS-NDC and would greatly reduce the burden on the center. However, transmission of all the array data to the IRIS-DMS with on-line archival and servicing of user requests would be excessively costly. Only moderate savings can be achieved by retaining segmented event windows on line, given the need to process the continuous data in order to obtain the segments. The probable high cost and redundancy of the massive archive makes this approach unattractive.

Another approach is for the ISMS-NDC to take the role of extracting specified event-windowed segments from the array data and either archiving these in a convenient retrievable form and directly servicing user requests for the segmented data, or passing the greatly reduced data set on to the IRIS-DMS for archival and distribution. Guidelines for the precise data windows could be established by the research community, with preliminary feedback to the panel indicating primary interest in data for global events with magnitudes above 4.0, although in regions of possible decoupling, events as small as 2.5 are of interest. This approach also limits servicing of user requests by the ISMS-NDC, but places restrictions on the data availability and requires that ISMS-NDC resources be invested in the production of a windowed data set that may not receive extensive use.

A third approach is to utilize the internal database management system of the ISMS-NDC to provide access to the array data. Here the major concerns are to minimize impact on the nuclear monitoring function and to protect the security of classified data bases at the ISMS-NDC. Having the ISMS-NDC service multiple user requests for the array data would require extensive software development and would require seismic data users to become familiar with a new data access system (in addition to the IRIS-DMS, USGS, and university data centers). The GSE, ARPA, USGS, FDSN, and IRIS have designed and tested a prototype of an automated data access system that could provide on demand requests for array data as well as service standing subscriptions for specific parameter windows. Such a system could be linked to USGS/ISMS-NDC or IRIS/ISMS-NDC interfaces to provide full access to the array data.

***The panel recommends that access to all continuous array data be made available by a system that exploits existing seismic data distribution capabilities.***

An interface to the ISMS-NDC data management system should be installed to allow on-demand requests for array data as well as standing subscriptions for specific parameter windows. If this is implemented, requests for array data would be submitted to IRIS or to the USGS in familiar formats like those used to obtain broadband or regional array data; an IRIS DMS/ISMS-NDC or USGS/ISMS-NDC interface would retrieve the specified data windows. This approach will (1) minimize the external user request workload for the ISMS-NDC, (2) maintain a centralized access point for broadband and array data for the seismic research community, and (3) retain access to the entire ISMS data stream. Restricting the number of users with access to the ISMS-NDC data base simplifies data security considerations for pathways to other data sets collected by NTM, and ensures that the system performance is not degraded by multiple-user access. The cost for setting up this interface should be borne primarily by the nuclear monitoring research agencies, as most current user data requirements involving short-period array data are motivated by nuclear monitoring issues, although other applications will grow. Efficient utilization of the array archive would be facilitated if 30-minute segments of data from array beams for events above a certain magnitude (say 4.0) were routinely provided to the research community via IRIS or the USGS. This would allow the research community to assess event signal quality and to make requests for only promising data. This could be in the form of a standing subscription request.

*The panel recommends that a prototype interface for accessing the array data be installed during the GSETT-3 experiment.*

The GSETT-3 test can provide a test of the ability of the NDC data management system to service both national verification functions and the external array data user community. Based on this experience, procedures for distributing the array data may have to be revised.

The precise way in which the U.S. ISMS-NDC is defined is a political issue, and there are inevitable tensions over agency roles. The panel's recommendations are directed essentially at how the ISMS-NDC will function, not at how it is internally funded. The data from the ISMS-IDC should have a direct pipeline into the national verification arena, which is likely to be at AFTAC. It is probable that the nonseismic data will come into the same location, making this the hub of the U.S. monitoring efforts. The panel believes the broadband data will be passed on to the USGS and IRIS, to be merged with other global network data. It is also possible that some users will want to go directly to the ISMS-NDC, even for broadband data. Thus from one perspective, the USGS and IRIS could be viewed as users of the ISMS-NDC data, while from another perspective they could be viewed as part of the archival and distribution function of the ISMS-NDC. This should not become a political flashpoint, as the ultimate objective is to ensure long-term general access to the data in a cost-efficient manner. There is great value in consolidating the broadband data in a single unified data base, with many years of on-line data.

The U.S. ISMS-NDC will receive extensive seismic event information from the ISMS-IDC. This information could be of substantial use for earthquake monitoring efforts.

*The panel recommends that the seismic event parameter information (arrival times and hypocentral parameters) and final bulletin from the ISMS be made available to the earthquake monitoring agencies rapidly via the ISMS-NDC.*

These parameter data should include event locations and phase information as well as other event parameters determined by the IDC. The current plans for the ISMS are for such information to be provided to the ISMS-NDC within 48 hours, and the information should immediately be made available to the earthquake monitoring agencies. The ISC produces the definitive global seismicity catalog for use by the international seismic research community, building upon the EDR of the USGS.

*The panel recommends that the seismic event parameter information (arrival times and hypocentral parameters) produced at the ISMS-IDC be transmitted by electronic means to the ISC, to be incorporated in their final bulletin preparation.*

This recommendation should involve negligible extra cost for the ISMS-IDC operation, but would almost certainly add costs to the ISC operation. Additional seismic information collected by the ISMS-IDC involving parameter data (arrival times and event locations) from regional arrays should also be made available to the earthquake monitoring program and the ISC. It is important to note that the high-quality earthquake bulletins prepared by the USGS and ISC will be valuable for the nuclear monitoring operations for improving station corrections, for providing background activity levels, and for providing a basis for comparison with the IDC/NDC operations.

*The panel recommends that distribution of the parameter data to earthquake monitoring and bulletin preparation agencies be initiated in the GSETT-3 experiment.*

The recommendations given above are designed to establish a U.S. infrastructure that meets the needs of both the nuclear test and earthquake monitoring communities in the short and long term. The ISMS data set will be very large, so it is desirable to minimize redundancy in archiving the data. It is important to recognize that much of the substantial infrastructure necessary for providing convenient user access to the archive already exists. The IRIS-DMS has proven an effective center for archiving and distributing broadband seismic data in response to tailored user requests. Rather than replicate this capability at the ISMS-NDC, it would be more efficient to transmit all the broadband data to this facility, along with necessary instrument calibration information. This can be synchronized with transmission of the broadband data to the earthquake and tsunami monitoring operations. Alternatively, the IDC data could be available in parallel to many users/data centers if appropriate satellite downlink protocols were established.

It should be noted that there is precedent for providing parallel access to seismic data streams in the envisioned manner. The Global Telemetered Seismic Network/Ancillary Seismic Network (GTSN/ASN) data collection and distribution to AFTAC and through the USGS and IRIS provides a functioning model for this type of interaction. By providing access to the continuous array data through the USGS/ISMS-NDC or an IRIS-DMS/ISMS-NDC interface, the entire ISMS data set will be available to the community, but the ISMS-NDC burden of servicing data requests will be minimized. Nothing would preclude the ISMS-NDC from also servicing requests for broadband data. The objective here is to provide guaranteed access to the data in a form that is as convenient as possible for the researchers. Such a structure will ensure that there is no deleterious effect on the nuclear monitoring operation while enabling optimal multiple use of the data. This integrated approach, involving open access to all unclassified seismic data, offers the first opportunity to work toward a rational U.S. National Seismological System servicing national needs.

# 5

# ISMS AND U.S. NATIONAL VERIFICATION RESEARCH AND DEVELOPMENT INFRASTRUCTURE

In this chapter, the third charge is considered, which concerns the research and development infrastructure required to support the proposed ISMS. This section presumes that the data quality and data access issues discussed in the preceding chapters will be adequately addressed; it focuses on the research and development infrastructure and knowledge transfer that will provide technical support for the ISMS and U.S. CTBT monitoring operations.

Monitoring a CTBT poses many unprecedented technical and scientific challenges, and there will be a continuing need for basic and applied research and advanced technology and automated systems development in all of the disciplines that contribute to the monitoring system (OTA, 1988). Given the imminent implementation of a prototype CTBT monitoring system, it is critical to have an integrated and reviewed program that carries out basic research, tests the results in operational settings, and implements useful, cost-effective advances in the operational system. This holds for both the ISMS and the U.S. monitoring systems.

While CTBT monitoring is intrinsically an arms-control issue, primary responsibility for seismological nuclear test monitoring and research has historically resided within the Department of Defense (DOD) and the Department of Energy (DOE). Given the current climate of shifting organizational roles, it is not clear which government organization is ultimately most suitable for overseeing CTBT monitoring efforts. It is assumed that the mission for monitoring research will continue to reside within DOD and DOE, with supporting activities by the USGS and seismological research community. It is important that if these agency roles change, the basic seismological research effort be maintained by those responsible for the functions of monitoring, verification, and hazard reporting. This chapter describes basic mechanisms by which participating agencies can implement an effective research and development program in seismology servicing the operational needs of the CTBT monitoring effort.

The actions recommended below will ensure that an integrated research and development system consisting of basic, applied and advanced development elements supports the CTBT monitoring efforts of both the ISMS and the U.S. nuclear monitoring systems. More effective coordination of the overall research program and more efficient transfer of technological advances into the operational regime will result from implementation of these recommendations. They will also support the development of personnel with appropriate expertise and capabilities.

## Introduction and Background

The panel was asked to consider the following charge:

*Research Feedback.* An important aspect of the GSE concept is that the system can evolve. This includes regular improvement of the processing capabilities (e.g., travel-time and amplitude path corrections, enhancement of phase identification and event location, and new processing techniques). What is the best way to implement promising basic and applied seismic research within the GSE system? To what stage must research be taken (e.g., publication, algorithms, or finished software) to most expeditiously and reliably implement it within the system? What are the long-term national research and development programs required to support the envisaged monitoring system?

The panel issued a broad request for feedback on research and development infrastructure issues. The responses highlight several weaknesses of the past and present infrastructure related to research and development for seismological monitoring of nuclear testing treaties. A wide range of issues surfaced, involving program guidance, funding issues, need for a research testbed, and basic structural problems within participating agencies. It also became clear that the issue of research in support of the GSE system or ISMS system is only part of the broader issue of how seismological research should be organized to service the U.S. national verification effort, particularly since it is not clear that the ISMS will have any event identification responsibilities. We now provide some background on the existing research and development infrastructure.

The research and development efforts in seismology that support nuclear test monitoring date back to the 1959 Berkner Panel report "The Need for Fundamental Research in Seismology". That report provided a rationale for fundamental research in seismology as an integral part of any successful treaty monitoring system. Many of the original arguments remain valid today. While the intervening 35 years have brought great advances in our knowledge of earth structure, global seismicity, seismic wave propagation, and characteristics of nuclear explosion and earthquake signals, the technical requirements for monitoring a sequence of nuclear testing treaties (the 1963 Limited Test-Ban Treaty, the 1968 Treaty on the Non-Proliferation of Nuclear Weapons, the 1974 Threshold Test-Ban Treaty, and the 1976 Peaceful Nuclear Explosion Treaty) have kept pace with, and even temporarily exceeded the seismological capabilities. We are on the threshold of another quantum jump in the required monitoring capabilities, as negotiations progress toward eventual signing and entry into force of a Comprehensive Test-Ban Treaty.

Many technical challenges confront the new CTBT monitoring effort, and continued seismological research is essential to ensure adequate U.S. national verification capabilities. Effective verification of a CTBT will require detection, location, and identification efforts using regional and teleseismic data from sources and stations distributed worldwide. The problems to be faced involve critical issues such as how to distinguish the small seismic vibrations from large quarry blasts from the vibrations produced by a small nuclear explosion. Given the many regions of the world for which there is little or no familiarity with the crustal effects on seismic waves, the accuracy of event locations and the confidence in event identifications will be unacceptable until research efforts calibrate each region. These efforts will require seismologically trained personnel, monitoring systems capable of processing large numbers of events, and new algorithms for detection, location, and identification tuned to specific regions of Earth. Availability of the personnel, algorithms, databases and systems that are needed requires a program that trains personnel and provides an orderly transition for concepts and algorithms from their conception to implementation in the operational systems.

An integrated program that supports a continuum of efforts from basic research to the operating system will provide the United States with effective CTBT verification capabilities. The continuum includes the following categories.

- Basic research efforts that tend to be focused in the universities. These efforts train personnel, develop new theories and relevant concepts, carry out investigations in areas of interest, and provide useful data. Long-term research efforts on basic seismic wave propagation, source theory, and techniques to extract information from waveforms are concentrated in the basic research area.
- Applied research efforts that tend to be concentrated in the efforts of the private contractors, with some contributions from universities. These efforts look to both basic research and operational needs to determine their priorities. They develop the concepts identified by the basic research into functional algorithms and expand and organize the data bases begun there. They interact with the operating systems to develop

algorithms and carry out studies to address specific problems encountered by the operating system in the course of its day-to-day functions. Finally, they develop concepts for the operating system and develop elements that can be incorporated into it. The contractors are a major part of the employment environment that provides job opportunities for the students and post-doctoral fellows who have gained experience in the basic research efforts. Sustaining this job market should help to attract university researchers to this area.

- Advanced development research that tends to be concentrated in the contractors and the operating agencies. These efforts develop new operating systems that integrate hardware and software for data acquisition, communications, data archives, processing, analysis, and display.

Although past efforts spanned these functions, the current demand for functional CTBT monitoring systems tends to stress advanced development and applied research, and most research funding should be oriented in these areas. However, the long-term challenges of CTBT monitoring will require the development of new detection, location, and identification concepts and the availability of highly trained personnel. These assets will only be available if the research program includes a strong basic research element as part of the overall integrated program. Given the fact that many of the scientists trained in the seismic research programs of the 1960s are nearing retirement age, the panel is concerned that serious degradation of the basic research program will result in a lack of qualified, knowledgeable verification seismologists at the time that the CTBT enters into force. The panel acknowledges the fiscal considerations but is vehement in its support for the continuing need for a strong research effort.

Over the past several decades, the Department of Defense (DOD) has held the primary role in supporting basic seismic research and development for nuclear monitoring. Actual monitoring operations have been conducted by the Air Force Technical Applications Center (AFTAC). It is essential to sustain research, development, and transfer to the operations within DOD. In principle, the DOD nuclear monitoring research program supporting this operation has a standard DOD subdivision into basic research (the so-called 6.1 program), exploratory development research (6.2 program), and advanced development research (6.3 program) components. The DOD research effort is currently organized under the Air Force Office of Scientific Research (AFOSR) (6.1), the Air Force Phillips Laboratory (AFPL) (6.2), the Air Force Technical Applications Center (AFTAC) (6.3), and the Advanced Research Projects Agency (ARPA). In practice, AFOSR and ARPA have, at various times, had dominant cross-cutting roles that have combined the 6.1 and 6.2 functions.

The DOD 6.1 basic research program, currently administered by AFOSR, has an annual funding level of about $4 million for external research. This program supports fundamental relevant research investigations, with a current emphasis on small-event discrimination. The 6.2 exploratory development program has the goals of identifying and extending promising research results from the 6.1 (AFOSR) program and transitioning these to AFTAC, the operational client. The 6.2 exploratory development program was sponsored by ARPA in the past (with administrative support primarily through AFPL), but as of FY95, the responsibility for this effort was transferred to DOE. Currently, a scaled-back 6.2 program resides at AFPL, with no direct funding for external support. In 1995, AFPL began to administer some of the exploratory development projects funded by AFTAC and DOE. A 6.3 program has been administered by AFTAC and is intended to develop advanced systems and mature technologies for the operational environment. For the next two years, AFTAC will be supporting about $3.2 million/yr of external research funding for efforts in the 6.2 area, with a corresponding reduction of the normal 6.3 effort that AFTAC would support. The total AFTAC CTBT verification budget for operations, research and development in seismology is $24 million for FY95 (DOD, 1994).

In the past, ARPA was a primary supporter of all levels of seismological research, with an increasing emphasis over the last 10 years or so on applied and advanced developmental research. ARPA has been involved in developing seismic arrays for use in nuclear monitoring, but starting in FY95, responsibility for deploying and operating new arrays was transferred to AFTAC. Currently, ARPA's seismological effort is focused on development of a prototype ISMS International Data Center, with a total ARPA CTBT verification research and operations budget of $13.8 million for FY94 (DOD, 1994). Over the next two years, ARPA may provide some funding to the 6.2 research program administered by AFPL, using funds provided to bridge a phase-out of ARPA research and development in nuclear monitoring seismology.

Beginning in FY95, the Department of Energy (DOE) was assigned the mission "to carry out research and development necessary to provide U.S. government agencies responsible for monitoring and/or verifying compliance with a CTBT with technologies, algorithms, hardware, and software for integrated systems to detect, locate, identify, and characterize nuclear explosions at the thresholds and confidence levels that meet U.S. requirements in a cost-effective manner" (DOE, 1994). Much of this mission had previously resided with ARPA, but the establishment of the DOE program was not intended to preclude continued DOD research efforts, and indeed most of DOD's 6.1, 6.2, 6.3 and DOE's programs are now closely coordinated. The external funding provided by the DOE program is about $4 million/yr, with the program (DOE, 1994) being strongly oriented along the lines of exploratory and advanced development research. At DOE's request, AFPL administers the DOE external grants in this program.

The annual DOE budget for CTBT monitoring research in areas other than space is about $24 million (DOD, 1994). This program builds on long-standing internal research programs on test-ban verification at the DOE national laboratories.

It is obvious that DOD and DOE agencies engaged in CTBT monitoring activities will do their work with an emphasis on the U.S. needs to monitor the rest of the world. A somewhat separate issue is the importance of U.S. help in building up international programs for CTBT monitoring as well. This activity is of specific interest to policy agencies such as the U.S. Department of State and the U.S. Arms Control and Disarmament Agency, which is in charge of coordination with other U.S. agencies and with the conduct of the negotiations of the Conference on Disarmament in Geneva.

The National Science Foundation (NSF) and the U. S. Geological Survey (USGS) fund basic research and instrument deployments in seismology that have substantial potential relevance to the nuclear monitoring arena, but these agencies do not have explicit missions regarding nuclear monitoring operations. Development of earth models, basic research on earthquake source physics, and studies of regional tectonics are a few of their areas that intersect the nuclear monitoring arena. There are plans to incorporate many of the NSF/USGS seismic stations in the ISMS, as well. The USGS is helping the U.S. ISMS-NDC to provide data from stations located in the United States to the ISMS-IDC. The United States must have the capability of prompt and detailed reporting for seismic sources on U.S. territory (earthquakes, routine and unusual mine blasting to the extent that the consequent signals are detected by the ISMS-IDC, accidents with explosions, and large-scale explosions in military programs). The USGS is the existing U.S. agency best able to document seismic activity in U.S. territory, as its seismological monitoring effort involves a great number of stations in North America. The U.S. Bureau of Mines may also help in this effort.

The research efforts mentioned above have had varying degrees of coordination over time. For 17 years AFPL, AFOSR, and ARPA programs have held an annual Seismic Research Symposium that has brought together the basic and applied research communities. These are distributed across universities, private companies, and federal agencies. In 1994, the AFPL and AFOSR Seismic Research Symposium and a meeting of ARPA contractors were held separately. In 1995 ARPA again organized a separate meeting in May for its contractors and European leaders of the GSE, while AFTAC, AFPL, AFOSR and DOE met jointly in September. These separate meetings serve to exacerbate the growing rift between the basic and applied programs. There is a general perception that communications between the operations and research communities have not been effective. A Seismic Review Panel provides advice to AFTAC, but is not charged with coordinating the research needs of the operational environment and the basic research programs. The challenge of monitoring a CTBT will heighten the need for efficient development and transfer of technologies from basic research to operational capability.

The technical difficulties and research program strategies for monitoring a CTBT have been thoroughly addressed in several recent documents (AFTAC, 1994; Blandford et al., 1992; DOD, 1994; DOE, 1994; van der Vink et al., 1994). Appendix B tabulates some of the seismological research efforts that are being developed and conducted in the DOE research program. In brief, the major change from past monitoring efforts is that CTBT monitoring will require global identification of events down to low seismic magnitudes, whereas previous Threshold Test-Ban Treaty monitoring emphasized yield estimation for large explosions at Soviet and Chinese test sites. The need to detect, locate, and identify relatively low-magnitude events mandates the use of short-period seismic wave energy recorded at arrays and the use of high-frequency data recorded at regional (< 1200 km) distances.

Although research on regional seismic wave propagation is a rapidly developing area, fundamental issues remain regarding event discrimination with regional signals that are the topic of many current investigations. A direct consequence of the need to analyze small-event signals is that very large numbers of events, perhaps 100 to 300, must be analyzed on a daily basis. This requires enhanced automation and computer assisted decision-making in the operational environment. Nuclear tests must be discriminated from large quarry blasts, earthquakes, and rock bursts. The number and characteristics of these events pose major challenges to seismic analysis procedures. The very mature methodologies of yield estimation for large explosions developed for monitoring the Threshold Test-Ban Treaty are not directly relevant to monitoring a CTBT. The technical challenges are such that continued basic and applied research is vital to the long-term success of a CTBT monitoring effort.

## U.S. Research and Development Infrastructure

The effective development and transfer of new seismological advances into the CTBT monitoring operation require a well-coordinated, effective research and development infrastructure. We will first focus on the large-scale characteristics of the necessary infrastructure, and then provide specific suggestions as to how to achieve effective research and development support of the operational effort. The exact structure of the U.S. ISMS National Data Center is not yet worked out. For the purposes of this report we assume that the primary responsibility for U.S. nuclear test monitoring operations will continue to reside with the Air Force, while DOE will continue to have extensive seismic and nonseismic research and development activities relevant to CTBT monitoring and the USGS will assist in providing data on seismic sources on U.S. territory. Although, the panel recommendations are given in the context of continued DOD and DOE involvement, the basic rationale underlying the recommendations is not agency specific. In addition, the recommendations of this section are not restricted to the ISMS, but are perhaps even more relevant to the U.S. monitoring system.

In order to meet the challenges of CTBT verification, it is critical to establish stable and sufficient levels of funding for the overall CTBT monitoring research effort. Upheavals and annual crises in the DOD funding sources (ARPA, AFOSR, AFPL, and AFTAC) over the past 5 years have reduced the number of Ph.D. students in seismology who are working on relevant problems and have caused many leading university seismologists and consulting groups to back away from the program. While the present situation is somewhat more stable, continued involvement of the university and contractor resources requires that the agencies (DOD and DOE) funding research in this area commit to funding relevant efforts for clearly defined times. This commitment is vital for the training of a new generation of seismologists committed to nuclear monitoring research and operations, as well as providing employment opportunities in the contracting companies. It will also provide the entry point for new computer technologies, largely driven by university research, to be brought to bear on treaty monitoring applications. These will be important for CTBT monitoring in a resource-limited environment.

The various agencies that are contributing to CTBT monitoring efforts all recognize the technical demands that these efforts will place on seismological capabilities. While seismological methods are quite advanced in general, development of the necessary global understanding of regional seismic wave propagation, the fundamental nature of regional wave discriminants, and regional earthquake and quarry blast source characteristics requires resolution of fundamental research issues. While there must be a substantial emphasis in the overall research program on applied research and advanced systems development, there will be a continuing, probably long-term need for a fundamental research program relevant to the nuclear monitoring effort. This will ensure that technological and theoretical advances in the rapidly progressing field of seismology are brought to bear on nuclear monitoring issues. Operations under a CTBT will require highly trained analysts and decision makers. The universities can make important contributions both in training and in appropriate research. The contracting companies that participate in the applied research and advanced systems development programs will provide one of the major job markets for university-trained seismologists.

Assuming that DOD will continue to have a major operational CTBT monitoring function, the panel feels that it is critical that DOD sustain a research and development program in seismology with an integrated basic, applied, and advanced development effort. This program should be based on the standard DOD hierarchical structure of 6.1,

6.2, and 6.3 research efforts, each of which has a natural home within the existing Air Force structure. Any departure from the recognized DOD hierarchy weakens the overall effort and can lead to instability when uncoordinated actions are taken in different units. The DOE research and development program and the DOD effort should be closely coordinated to avoid redundancy. The coexistence of DOD and DOE research and development programs will best exploit the distributed expertise within the agencies, and as long as there is effective coordination, the combined effort should service the monitoring needs.

The existing DOD basic research (6.1) program administered by AFOSR contributes fundamental research in regional wave propagation, event location, and source discrimination. In the short-term it would be valuable to augment this program to enable additional exploration of the basic physics underlying the regional seismic wave discriminants that are being proposed for the operational environment. Long-term stability of this program should be established in recognition of the heightened challenges of CTBT monitoring and the difficulties of anticipating what research directions will advance operational capabilities. (Applied research programs tend to lack the innovation and breakthrough discoveries that are common in basic research efforts.) However, there is concern within the operational effort at AFTAC regarding the effectiveness of the 6.1 program. It appears that there is a need to enhance communications across the DOD research hierarchy to ensure that the basic research program does emphasize the problems relevant to the operations.

***The panel recommends that the basic research (6.1) program in seismology, currently administered by the AFOSR, should be sustained and expanded to maintain an influx of researchers and fundamental research on long-term problems associated with seismological monitoring of a CTBT.***

Within the standard DOD framework for research, the promising developments of a basic research (6.1) program are further developed in an exploratory development, or applied research (6.2), program. Typically, this involves a greater percentage of private company contractors than university contractors. The present situation for DOD seismic research is not optimal, in that AFPL, the Air Force branch explicitly identified as having responsibility for applied seismology research, has no funding of its own for external research in this area. This is a consequence of the past few years of unsettled and politicized budgets. External funding in the 6.2 program of $2 million/yr has evaporated. In the standard DOD model, AFPL would conduct research that directly follows up on the AFOSR basic research program. At present, AFPL does administer contracts for directed research efforts sponsored by DOE, AFTAC, and (ephemerally) ARPA, but these are not designed to ensure effective technology transfer from the

AFOSR program, as the funding sources are quite focused on enhancing short-term operational capabilities.

While the existing directed research effort appears to be functioning very well, the AFOSR program is undesirably isolated from the operational environment. It is possible to improve the current situation by providing stable funds for external research to AFPL, with the explicit charge of effectuating the desired technology transfer, as long as strong communications are put in place to bridge the diverse components of the DOD research effort. The 6.2 funding should be coordinated with and should complement the existing directed research programs sponsored by DOE and AFTAC. The establishment of stable 6.2 funding would remove the need for AFTAC to divert some of its advanced development resources, as is occurring at present, and a stronger 6.3 program could be established.

***The panel recommends that the Air Force exploratory development research (6.2) program in seismology, currently administered by the AFPL, should be provided with a stable base for external funding to enable effective development and transfer of promising research and technologies from the Air Force basic research program to the Air Force operational environment. AFPL should continue to administer the external DOE program, as participation in this joint effort provides a natural mechanism for the various elements to coordinate their research effort.***

The CTBT monitoring emphasis on regional waves in diverse areas of the world requires extensive application of basic seismological techniques to characterize wave propagation and source characteristics in many regions. Sufficient funding to conduct field studies and calibration efforts to perform systematic regional characterization, to pursue new basic research developments, and to integrate the results into a standard format and/or database should be provided to the applied research effort.

AFTAC has supported advanced development research (6.3) for many years, and this continues to make excellent sense as long as AFTAC is the operational arm of the nuclear monitoring effort.

***The panel recommends that the advanced development research (6.3) programs in seismology, currently administered by the Air Force Technical Applications Center, should be sustained.***

The large seismic databases to be collected by the ISMS and the requirement of analyzing vast numbers of small events creates large computational demands for the CTBT monitoring system. For several years, ARPA has overseen the development of advanced computational platforms that can efficiently perform the data analysis. This effort should be sustained. Exploration of new computer technologies and intelligent computer systems is within the purview of ARPA and is an element of the DOE program. It is clear that communications technologies are going to continue to advance at a remarkable rate and that entirely new monitoring capabilities will emerge in the future. A natural, and important, role for ARPA is to be looking forward to the next generation monitoring system, once it is no longer operating the prototype IDC. New strategies for event location and identification will be enabled as the number of seismic data streams increases, and it is realistic to imagine real-time access to hundreds, or perhaps thousands, of seismic stations from global regional networks as well as arrays and broadband stations. The communications, data processing, and data archival challenges will be significant. Meeting them will require advanced computer technologies.

*The panel recommends that the development of the prototype ISMS International Data Center, currently being performed by ARPA, should be sustained, the exploration of new computer technologies and intelligent computer systems by ARPA and DOE be coordinated, and the results incorporated in the U.S. monitoring system, to the extent appropriate.*

One of the major challenges in the DOD research structure is to ensure that it actually functions, with relevant projects funded under the 6.1 program being picked up by the 6.2 program, and, if promising, passed to the 6.3 program. At present no formal overview of all levels exists. Conventional mechanisms for research documentation, such as reports and meeting presentations need to be reconsidered to enhance the technology transfer from one program to the next. Below, we discuss the concept of a research test bed, which might provide a more effective means for moving research developments into the operational system.

*The panel recommends that the Air Force basic research (6.1), exploratory development (6.2), and advanced development (6.3) programs should be coordinated by a group or organization that is aware of the monitoring requirements, the operational needs, and current and past research efforts.*

In response to its recently acquired responsibility for providing research and development efforts to support U.S. monitoring agencies, the DOE has developed a detailed, focused research plan (DOE, 1994) for its internal and external CTBT monitoring research program. This program draws upon the DOE national laboratories and supports applied research and some advanced development efforts at universities and in the private sector. It provides limited support to basic research. DOE and AFTAC are engaged in a continuing exchange in order to minimize impediments to technology transfer across agency boundaries. This interaction needs to be maintained and expanded. The program appears to be functioning well and has an appropriate level of support.

***The panel recommends that the directed research and advanced development research programs in seismology currently administered by the Department of Energy should be sustained. A knowledgeable, responsible advisory mechanism should oversee the combined DOD/DOE research effort to ensure relevance and continued coordination of the programs.***

Given a stable funding structure and a closely coordinated basic, applied, and advanced development research program, it is still of major importance to that researchers be well informed of the concerns of the monitoring environment. This has not been achieved very successfully in the past. There have been, and will continue to be, issues associated with restricted access to classified data sources and procedures. For example, while the ISMS will be an open system, with unclassified data, the U.S. monitoring effort will involve additional classified data sources. Private companies have commonly performed classified research for ARPA and AFTAC, and it is appropriate for such activities to continue as needed. Universities, and many private companies, perform unclassified research. It is important that classification not be an unnecessary impediment to performance of relevant research. To the extent consistent with national security issues, the research community funded by DOD and DOE programs should be provided with information about the operational methodologies and the operating system, along with access to relevant data (such as the ISMS data). This is critical for focusing the research efforts on relevant issues. Where security issues intervene on critical topics, appropriate clearances should be provided to promising researchers.

Feedback from the operational system to the research environment could improve relevant problem solving. For example, providing lists of problem events and an indication of the nature of the ambiguity could focus research on the most important operational problems. Other forms of outreach, such as newsletters from the monitoring community and site visits to both funded and unfunded seismic research universities and companies, could focus research attention on relevant problems. Clearly, national security concerns impose some limits on communication, but there is no question that communications can be improved. The annual Seismic Research Symposium provides a good opportunity for researchers, program managers, and operations personnel to define and discuss the specific monitoring problem that each applied research effort addresses, its context, and long-term concerns that might be addressed by basic research. The fragmentation of this symposium in 1994 and 1995 should be reversed, as this undermines communication across the participating research levels. Continuing the current Broad Agency Announcement (BAA) process is important to attracting a wide range of researchers and new technological approaches to the nuclear monitoring programs.

A valuable feedback to the monitoring effort would be a systematic comparison of results from the U.S. monitoring system with the results of earthquake monitoring operations. This would involve comparison of event catalogs and the associated location and size estimates. Earthquake monitoring information such as fault mechanisms and source characteristics may also be useful for calibration efforts in the monitoring operation. This feedback is an integral part of assessing the system performance and is a means for validating the operational procedures. Numerous organizations have contributed, and will continue to contribute, to our understanding of regional seismicity independent of the nuclear monitoring community. The results of these efforts should be exploited in nuclear monitoring activities.

*The panel recommends that improved communication between the DOD operational environment and researchers in the basic and applied programs be fostered. Release of information about operational methodologies and procedures, lists of problem events, and comparisons of seismic bulletins from different communities are among the activities that could enhance responsiveness of the research community to the operational requirements. Communication across the various elements of the monitoring and research communities should be fostered. Symposia, site visits, and advisory panels should be part of this communications effort. Focused experiments, involving broad communities, should be conducted to concentrate effort on important issues.*

One of the most difficult aspects of all research program infrastructures is the interface with the operational environment. This holds equally true for the ISMS and the U.S. monitoring operations. As research ideas develop and mature in the basic and applied research programs, it is critical that they be tailored to the operational needs. This has been performed inefficiently in the past, primarily by AFTAC assigning some of its limited number of internal personnel to implement technologies emerging from the ARPA and AFOSR programs. Given the advanced computer environment required for the CTBT monitoring effort, it is important for researchers to be able to develop and test their research products in an unclassified test system. To the extent possible, consistent with national security considerations, a readily accessible test system should have the same (unclassified) data bases, data streams, and operating system as the operational environment. Thorough documentation of the software interfaces for the system should be provided so that new procedures can be interfaced and examined under realistic conditions.

The establishment of a test bed system should be accompanied by a ground-truth data base for assessing the performance of new methods and for deciding which approaches should actually be incorporated into the operating system. Research intended for the operational system should be documented, preferably with open publications, and the statistical performance on the ground-truth data base should be established. This evaluation should include information about false alarms and the probabilities of missed violations. Software that has been exercised in the test bed should be readily exported to the operational environment once it has been thoroughly validated. A test bed of this type will be an essential element of the effort to incorporate the most useful developments from the basic and applied research efforts. It will free the operations personnel from maintaining the system, while providing a platform for them to interact with the external research program.

***The panel recommends that, to the extent possible, consistent with national security considerations, an unclassified experimental test bed facility that replicates the basic U.S. and ISMS analysis procedures be established to enable new developments to be tested in a realistic environment, enhancing transfer of applied research results into the operational systems. Ground-truth data bases should be included to determine performance of new methods. Regionalized data bases relevant to detection and discrimination efforts should be established and made available to the research programs through the test bed facility.***

A special event data base should be maintained and made available to all researchers to eliminate laborious "rediscovery" of valuable data. This data base could be organized on a regional basis along the lines of the research program regionalization. This would include data for historical explosions, ground-truth events, and compilations of useful signals assembled by contractors in the course of their funded research. This data base could be made available through the test bed described above, or through an agency capable of maintaining and providing access to an expanding data base of this type. AFTAC may find it useful to maintain a data base with classified/restricted data for access by researchers with clearances.

***The panel recommends that a research data base of important seismic recordings be assembled and maintained.***

There are many research problems in seismology that are relevant to earthquake monitoring as well as nuclear monitoring operations. These include event catalog determination, with attendant technical issues such as improved association algorithms, multiple-phase location procedures, and event location in heterogeneous models. Interagency working groups bridging the earthquake and nuclear test monitoring agencies should help to coordinate and foster research with dual applications. This could help to activate relevant research activities in refined catalog preparation.

***The panel recommends that major research efforts that have potential benefits for both nuclear test and earthquake monitoring, such as enhanced association algorithms, new regional event location procedures, and event location procedures***

*in three-dimensional models, should be coordinated through interagency working groups.*

*The panel recommends that a program that supports postdoctoral fellows and visiting researchers should be established at both the International Data Center and the U.S. National Data Center. This will enhance communication and research coordination between the operational and research environments.*

The primary focus of the above recommendations has been the U.S. research program, but there is clearly valuable research conducted internationally as well. Historically, a significant portion of this has actually been funded through the ARPA program, and maintaining that program will sustain some of the international effort. Annual seismological symposia have provided a useful forum for interaction between U.S. researchers and their international colleagues, and that would be sustained by the foregoing recommendations. Presumably, some international research activity will be associated with different national data centers, and technological advances will feed back to the ISMS via those centers. International collaborations on field deployments that characterize regional crustal structures and seismic wave-propagation characteristics in different regions provide a natural means of mutually enhancing treaty verification capabilities, and such efforts should be supported by the DOD 6.2 and DOE programs.

This chapter has outlined a long-term research and development infrastructure to support the operational system that will monitor a CTBT. A balanced, well-organized program involving relevant basic research, applied research, and advanced development research is the logical approach. A guiding principle for a stable research effort is that the program be optimized within existing agency hierarchies, avoiding any gaps in the structure that may impede technology transfer. To ensure that relevant research is conducted and brought to operational capabilities, enhanced lines of feedback from the operational system and knowledgeable advisory panels need to be established, along with a realistic research test bed with ground-truth and relevant data bases. Implementing the recommendations of this chapter will ensure that CTBT monitoring efforts continue to have the critical influx of research innovations and technical developments vital to an effective monitoring operation. It should also be noted that many international research activities on nuclear monitoring problems are directly supported by the DOD. While this chapter has emphasized the needs of the U.S. program, the international efforts will be similarly strengthened if the recommendations are implemented.

# REFERENCES

Air Force Technical Applications Center (1994). *AFTAC Subsurface Research Plan,* Air Force Technical Applications Center.

Berkner, L. V., H. Benioff, H. A. Bethe, W. M. Ewing, J. Gerrard, D. T. Griggs, J. H. Hamilton, J. P. Molnar, W. H. Munk, J. E. Oliver, F. Press, C. F. Romney, K. Street, and J. W. Tukey (1959). *The Need for Fundamental Research in Seismology,* Department of State.

Blandford, R., A. Dainty, R. Lacoss, R. Maxion, A. Ryall, B. Stump, C. Thurber, and T. Wallace (1992). *Report on the DARPA Seismic Identification Workshop,* Department of Defense.

Department of Defense, Report to Congress (1994). *The Department's Plans to Develop Advanced Technologies for Monitoring a Comprehensive Test-Ban Treaty (CTBT),* Department of Defense.

Department of Energy (1994). *Comprehensive Test-Ban Treaty Research and Development FY95-96 Program Plan,* Department of Energy, DOE/NN-0003, November 1994.

Heaton, T. H., D. L. Anderson, W. J. Arabasz, R. Buland, W. L. Ellsworth, S. H. Hartzell, T. Lay, and P. Spudich (1989). *National Seismic System Science Plan,* U.S. Geological Survey Circular 1031.

National Research Council (1991). *Real-Time Earthquake Monitoring,* Committee on Seismology, Board on Earth Sciences and Resources, National Academy Press, Washington D.C., 52 pp.

Office of Technology Assessment (1988). *Seismic Verification of Nuclear Testing Treaties,* Congress of the United States, OTA-ISC-361. U.S. Government Printing Office, Washington, D.C.

Ringdal, F. (1985). Study of Magnitudes, Seismicity, and Earthquake Detectability Using a Global Network, In *The VELA Program, a 25 Year Review of Basic Research,* ed. Ann Kerr, Executive Graphic Services, p.620.

van der Vink, G., D. W. Simpson, C. B. Hennet, J. Park, and T. Wallace (1994). *Nuclear Testing and Nonproliferation: The Role of Seismology in Deterring The Development of Nuclear Weapons,* The IRIS Consortium.

# APPENDIX A

# CHARGE TO THE PANEL

In 1994, the National Research Council convened the Panel on Seismological Research Requirements for a Comprehensive Test-Ban Monitoring System to conduct a study requested by the Advanced Research Projects Agency (ARPA). The panel was to examine issues associated with establishing an International Seismic Monitoring System (ISMS) for verifying a Comprehensive Test-Ban Treaty (CTBT). Negotiations toward such a treaty are currently underway within the Conference on Disarmament (CD), with prototype versions of the ISMS being explored in a series of technical tests organized by the Group of Scientific Experts (GSE).

The CTBT monitoring system being considered within the CD includes the acquisition and processing of data from high-quality stations and provision of the data to participating states to assist them in their national verification functions. ARPA has requested advice on how the data from the CTBT monitoring system might best benefit the broader seismological community. The NRC panel has been charged with considering specific data characteristics desired by the broad seismological community, procedures for providing general access to the ISMS data, and the nature of a research infrastructure that could best support CTBT monitoring. Many of the same considerations apply to the U.S. infrastructure for CTBT monitoring.

The panel's task is summarized in three charges:

*Data Characteristics.* The Group of Scientific Experts (GSE) has written draft requirements for an ISMS-standard station that specify characteristics such as sample rate, passband, dynamic range, and sensitivity. They have also proposed a Primary Station Network configuration and rough requirements for signal detection, parameter extraction, and event location. What types of data (raw and/or processed) are sought by the seismological community for use in test-ban monitoring research and in other types of basic research?

*Data Access.* The GSE has specified that all authorized users (most likely, the ISMS National Data Center in each participating country) have prompt electronic access (perhaps through the ISMS International Data Center) to all raw and processed data. What kind of access would best satisfy the requirements of other operational groups (e.g., for earthquake hazards and tsunami warning)? How should the data be organized (e.g., by region, station, time period; continuous vs. event segments) and made available (e.g., access time scales—minutes or days; and media—electronic or optical)?

*Research Feedback.* An important aspect of the GSE concept is that the system can evolve. This includes regular improvement of the processing capabilities (e.g., travel-time and amplitude path corrections, enhancement of phase identification and event location, and new processing techniques). What is the best way to implement promising basic and applied seismic research within the GSE system? To what stage must research be taken (e.g., publication, algorithms, or finished software) to most expeditiously and reliably implement it into the system? What are the long-term national research and development programs required to support the envisaged monitoring system?

# APPENDIX B

# RESEARCH TOPICS FOR CTBT SEISMIC MOINTORING

This appendix identifies some of the many topics requiring continued seismological research in support of the treaty verification needs of the United States. Several research planning documents have been produced by AFTAC, ARPA, and DOE that emphasize the focused needs of applied research efforts. The DOE summary of research needs is reproduced in the next few pages.

Somewhat less consideration has been given to the objectives that should guide the basic and applied research programs in seismology that are now being managed by the Air Force Office of Scientific Research and the Air Force Phillips Laboratory. These organizations are in the process of commissioning a study by the National Research Council to help develop a plan for relevant basic research. While the detailed plan is not yet available, it is clear that it will identify some long-standing priority areas, such as improved theoretical and computational ability to model seismic waves in three-dimensional heterogeneous media; improved theory for excitation of seismic waves from diverse sources such as quarry blasts, chemical explosions, nuclear explosions and earthquakes; and new methods for characterizing the wave propagation effects of diverse geological environments, along with the effects on seismic event location and identification. Some of these issues are considered in the research plans for the applied and advanced development programs, but they require a longer-term approach than is characteristic of the latter programs. The basic research effort is also essential for drawing well-trained seismologists into the arena of treaty monitoring issues, to ensure a long-term supply of expertise required for the long-term task of reliably monitoring a CTBT.

The summary of the DOE Seismic Monitoring Research Plan (pages A3-A5; DOE, 1994) is reproduced here as an illustration of the types of research that must be sustained for the U.S. nuclear monitoring effort.

## Seismic Monitoring Research

*Goal:* The seismic monitoring research element's goal is to provide improvements in the seismic monitoring capabilities, primarily in regional location and identification and, to a lesser extent, in detection and characterization, to meet U.S. national requirements for CTBT monitoring. Improvements in all these functions will be made in the context of evolutionary upgrades to the prototype U.S. National Data Center (NDC). To the extent appropriate, these improvements will also be incorporated into the prototype International Data Center (IDC) being developed by the Advanced Research Projects Agency (ARPA) for use in the GSETT-3 experiment planned by the Group of Scientific Experts (GSE).

*Products:* The seismic monitoring program element will provide methodologies that define and improve the monitoring performance in high-interest regions and, to a lesser extent, the remainder of the globe; proven, documented algorithms for accurate event detection, location, identification ,and characterization; a basic understanding of the factors that control the performance of the algorithms so that they can be tailored to specific sites and regions; and an overview of the monitoring challenges posed by conventional explosions and the measures that can be taken to address these challenges. Supporting data bases, raw input information, procedures, and reports will accompany the final versions of the regional characterizations to AFTAC. The information that is acquired and the algorithms that will be developed will be applied (due to budget constraints) to two regions of interest only: southern and central Asia and the Mid-East/North Africa.

*Approach:* Event detection, location, identification, and characterization functional elements of the CTBT monitoring problem have common requirements for seismic data and regional characterization information. Task S1 is intended to provide regional geophysical and geological information about the Mid-East/North Africa and southern and central Asia that can be acquired from existing sources. Tasks S2 and S3 address the detection and location capabilities, respectively, in these regions. Task S4 develops an empirical understanding of existing identification concepts (discriminants) by testing them on data from the Mid-East/North Africa and southern and central Asia and quantifying their performance. Tasks S5 and S6 are efforts to understand the physical basis for location and discrimination, respectively, in order to develop methods that can be transported from one region to another. Task S7 defines and executes field studies to obtain significant new information or to resolve critical location and discrimination issues. Task S8 integrates the various elements across the projects in a comprehensive report.

## Task Overview

*Task S1.    Regional Characterization*

The goal of this task is to provide geological and geophysical information for the regions of high interest for use by Tasks S2-S6. Sources of natural and man-made seismicity and cultural noise will be identified and characterized. This information will be acquired from research either in the region of interest, including possible calibration experiments conducted under Task S7, or from technical contacts in the region and from seismic monitoring. It will be synthesized into reports on and data bases of velocity structures, travel-time curves, regional characterization of wave propagation, attenuation characteristics, and evasion assessments.

*Task S2.    Develop Detection, Phase Identification, and Event Association (DPIEA) Techniques*

The goal of this task is to develop new and/or improved regionally dependent algorithms for detection, phase identification, and event association in the Mid-East/North Africa and southern and central Asia regions.

*Task S3.    Develop Empirical Location (Epicenter and Depth) Techniques*

The goal of this task is to develop improved epicenter and depth estimates. These are likely to depend upon the properties of the specific regions. Significant improvements in epicenter location capability will benefit all aspects of treaty verification. More precise locations would greatly reduce the effort required in an on-site inspection. Event identification would benefit from improved depth estimates.

*Task S4.    Develop Empirical Discriminants in Areas of Interest*

The goal of this task is to test discriminants and determine the performance of existing and potentially useful regional ones in the southern and central Asia and Mid-East/North Africa regions. Both individual discriminants and combinations of discriminants will be studied.

*Task S5.*    *Develop Models for Regional Propagation and Event Location*

The goal of this task is to develop an understanding of the physical properties of both the regions under consideration and the recording network that controls the accuracy of the location and depth estimation efforts. The empirical results of tasks S2 and S3 will be used to develop a model that embodies the important propagation features observed in the region. The model will provide a basis for the validation, refinement, extension, or redefinition of existing location and depth estimation techniques and the development of new ones.

*Task S6.*    *Develop Models for Discriminants*

The goal of this task is to develop a physical understanding of the factors controlling the performance of existing event discrimination procedures. The results of Task S4 will be used to develop a model of the performance of discrimination techniques that could be generalized for all regions of interest. This task will provide a basis for the validation, refinement, extension, or redefinition of existing discrimination techniques, for the development of new techniques, and for the prediction of the performance of the discriminants in new regions.

*Task S7.*    *Perform Field Studies*

The modeling undertaken in tasks S5 and S6 will generate key questions regarding regional propagation and event identification that can be addressed only by field studies. Two types of field studies are envisioned: passive and active. In a passive field study, portable instrumentation would be deployed in the vicinity of targets of opportunity where seismic activity is anticipated. These could be earthquake aftershocks or routine blasting at mines or construction sites and other geologic settings of interest. This kind of field study is adequate for calibration of propagation models used in event location but would be inadequate for the explosion phenomenology development needed for event identification. In this latter case, source location and timing are critical. Therefore, an active experiment in which the experimentalists specify the time, location, and other source parameters is required. This task will design and implement both types of experiments, but only the active ones will satisfy the requirements of both the modeling aspects of location (S5) and identification (S6) simultaneously.

*Task S8.   Integrate Results*

This task integrates the results obtained in the various components of the seismic research project. For example, the magnitude of the mine monitoring problem for a given region will be summarized in a report drawing on the results of S1, S3, and S4.

# APPENDIX C

# ACRONYM LIST

| | |
|---|---|
| AFOSR | Air Force Office of Scientific Research |
| AFPL | Air Force Phillips Laboratory |
| AFTAC | Air Force Technical Application Center |
| ARPA | Advanced Research Project Agency |
| BAA | Broad Agency Announcement |
| CD | Conference on Disarmament |
| CDSN | Chinese Digital Seismic Network |
| CTBT | Comprehensive Test-Ban Treaty |
| DMS | Data Management System (cf. IRIS) |
| DOD | Department of Defense |
| DOE | Department of Energy |
| FDSN | Federation of Digital Seismographic Networks |
| FEMA | Federal Emergency Management Agency |
| GPS | Global Positioning System |
| GSE | Group of Scientific Experts |
| GSETT-3 | Group of Scientific Experts Technical Test #3 |
| GTSN/ASN | Global Telemetered Seismic Network/Ancillary Seismic Network |
| IDC | International Data Center |
| IRIS | Incorporated Research Institutions for Seismology |
| ISC | International Seismological Centre |
| ISMS | International Seismic Monitoring System |
| LNM | Peterson's Low Earth Noise Model |
| NDC | National Data Center |
| NEIC | National Earthquake Information Center |
| NIST | National Institute for Standards and Technology |
| NOAA | National Oceanographic Atmospheric Administration |
| NRC | National Research Council |
| NSF | National Science Foundation |
| NSN | National Seismic Network |

| | |
|---|---|
| NTM | National Technical Means |
| REB | Reviewed Event Bulletin |
| SAR | Synthetic Aperture Radar |
| SEED | Standard for Exchange of Earthquake Data |
| TCP/IP | Telecommunication Protocol/Information Protocol |
| USAEDS | United States Atomic Energy Detection Systems |
| USGS | United States Geological Survey |
| USNRC | United States Nuclear Regulatory Commission |
| VBB | Very Broad Band |
| VSP | Very Short Period |